Clarabel Gilman

Common animal forms

Third Edition

Clarabel Gilman

Common animal forms
Third Edition

ISBN/EAN: 9783337228804

Printed in Europe, USA, Canada, Australia, Japan

Cover: Foto ©berggeist007 / pixelio.de

More available books at **www.hansebooks.com**

LESSONS IN ZOOLOGY.

COMMON ANIMAL FORMS

BY
CLARABEL GILMAN

THIRD EDITION, REVISED.

BOSTON AND CHICAGO
NEW ENGLAND PUBLISHING COMPANY
1898

COPYRIGHT, 1892,
BY
CLARABEL GILMAN.

PREFACE.

This little book is the outcome of ten years' experience in teaching elementary science. It embodies the outlines of what I have found it wise to attempt with children, and is offered to teachers with the hope that it may prove suggestive and helpful. A special effort has been made to remove stumbling-blocks, by explaining points of structure that are likely to be puzzling, by giving minute directions for procuring and handling specimens, and by providing simple outline drawings that can be quickly copied upon the blackboard by one who has little artistic talent.

In general, the lessons are composed of two parts,— one in coarse type, consisting of short, clear statements of children's observations, frequently in their own words, with the facts that a teacher must sometimes supply in order to make a lesson complete in essential points; the other in finer type, containing directions for the teacher and additional facts, many of which the older children can be led to discover for themselves. In some of the lessons full illustrations have been given of the method of guiding pupils' observations by questions, which it has not been thought necessary to repeat in the study of every type, though the plan remains the same in all. If children are carefully taught to observe, describe, and compare in the manner outlined here, it is believed that they will have a good foundation either for the later scientific study of zoölogy, or for the intelligent observation of animal forms in nature.

I desire now to express my great indebtedness to Prof. Alpheus Hyatt and Messrs. D. C. Heath & Co. for the use of a large number of figures from the admirable series of "Science Guides" published by the latter. Their generosity has alone rendered it possible to illustrate the book so fully, especially in the lessons on insects.

The remainder of the cuts have been drawn from various sources.

CLARABEL GILMAN.

May 14, 1892.

CONTENTS.

	PAGE
THE SPONGE,	5
HYDRA,	11
SEA-ANEMONE,	15
CORALS,	18
STAR-FISH,	26
SEA URCHIN,	33
CLAM	41
OYSTER,	49
SNAIL,	54
EARTHWORM,	58
LOBSTER,	61
CRAYFISH AND CRAB,	69
HERMIT CRAB,	72
BEACH-FLEA,	75
SPIDER,	78
GRASSHOPPER,	83
CRICKET,	89
BEETLE,	92
DRAGON FLY,	96
BUG,	99
CICADA,	103
FLY,	106
BUTTERFLY,	110
MOTH,	113
BEE,	119
ANT,	126

AWARD FROM WORLD'S FAIR COMMISSION.

Lessons in Zoölogy received a diploma and bronze medal from the World's Fair Commission. The award reads as follows:

AWARD.

"*For a successful presentation of the subject in a manner suited to the comprehension of a child. The explanations and instructions as to handling specimens are such as would lead a beginner to lay a good foundation for future scientific study. The illustrations are simple and profuse, and can be reproduced on the blackboard.*"

LESSONS IN ZOÖLOGY.

THE SPONGE.

Lesson I.

For the teacher, a bath sponge with one or two large openings on the top; for each child, a straight wire hairpin and a slate sponge, are the things needful for this lesson. Each sponge may be cut vertically, almost to the base, through one of the large tubes, or vertical sections may be used with the whole sponges. The day before the lesson, each child should wash out his sponge and notice how it is changed by the water. Sponges should always be moist when studied. The hairpins are straightened out for use as probes.

The children have already learned the following things:

The hard, dry sponges took in water through all the little holes, and became soft and elastic. They are made of threads called fibres, whose ends project in little brush-like bundles on every side but one, and this side is darker and smoother than the others. There are many small holes in the sponge, and only a few large ones, or sometimes only one. One or two bright pupils notice that there are holes all through the sponge, and a large tube running straight down from the large opening.

Children will give some of these points spontaneously; others must be brought out by skillful questioning.

Now, being careful not to tear the fibres, we put the probes into the large openings, and trace the tubes (Fig. 1, *a*) into which they lead, almost to the base of the sponge. We put the probes into the small holes, and find small tubes (Fig. 1, *b*) leading from some of them to the large tubes; from others, cross-tubes (Fig. 1, *c*)

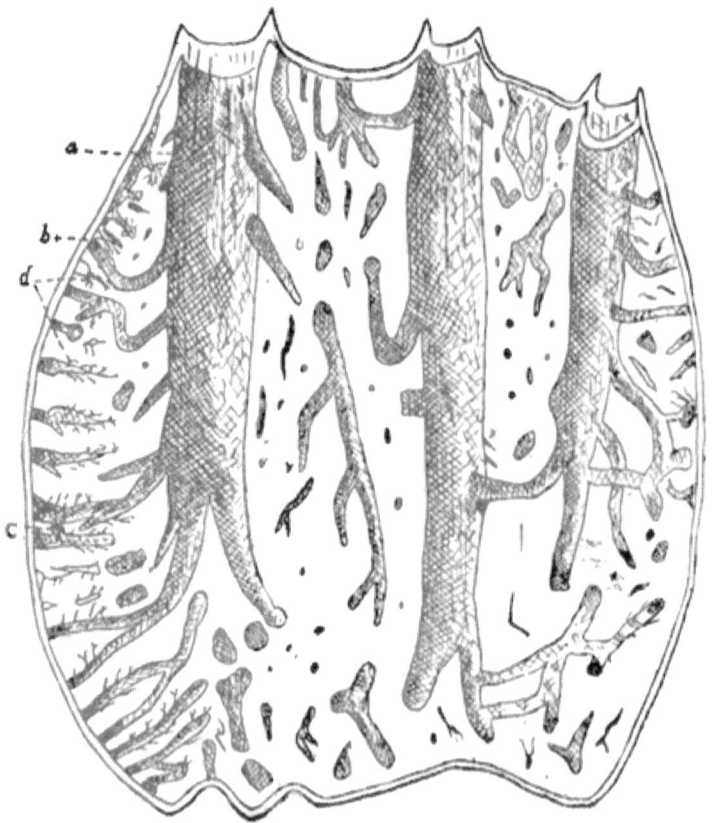

FIG. 1. Vertical section of glove sponge from Nassau, shown with the flesh.

connecting one small tube with another. Some of these connecting tubes in process of formation show plainly as channels on the surface, only partly covered in as yet by little bridges of fibres. Besides these tubes, there are others so small that we cannot trace them out (Fig. 1, *d*) passing in every direction through the mass of fibres.

A short talk about the home of the sponges ends the lesson.

Let us find Key West and Nassau on the map. These are the two principal markets for American sponges, which live in the Caribbean Sea and off the Florida coast. If we should visit Nassau, a boatman would take us out to the sponge fisheries. The water is very clear, and with a waterglass—a tube, or box, with a pane of glass at one end—which we press against the surface, we can see the bottom. Here and there on the coral rock, and contrasting with the brightly colored fishes and the brilliant hues of the sea-fans, are some dark masses fixed to the reef and sending out little jets of water from openings in the top. These are the sponges. We have in the boat a very long-handled fork, with three prongs, curved so that they will take a firm hold of the sponge, and with this our boatman pulls one off from the rock. Sometimes they are taken in a dredge, but the best sponges are brought up by divers. Our living sponge has a dark brownish or purplish flesh that covers all the fibres. After the sponges are killed by being exposed to the air for a day, they are thrown into pens made of stakes driven in shallow water, and left till the flesh decays. Then they are washed and trimmed, and sorted according to size, and afterward packed in bales and sent to New York or London to market.

This is true of American sponges. Mediterranean sponges, which are much finer and more expensive, receive more careful treatment.

Lesson II.

For this lesson every kind of sponge that the teacher can secure will be useful.

Review of Lesson I. : The sponge is a mass of elastic fibres. The edges of the fibres stand out on every side but one, which is smooth and dark. The sponge is full of tubes that open on the outside. There are four sets of tubes : large tubes, small tubes that lead from the surface to the large ones, cross tubes that connect these small tubes with one another, and microscopic tubes too small to be traced out. Our sponges come from the Caribbean Sea or the Florida coast, and were taken from the rocks with a curved fork or a dredge. Mediterranean sponges are brought up by divers. When alive they were covered with a dark-colored flesh, but the flesh has been removed, and only the fibres are left.

OUTLINE OF NEW WORK.

What was the use of the fibres? Not only to support the flesh, but also to protect the animal. They are made of a horny substance, and so tough that fishes very seldom

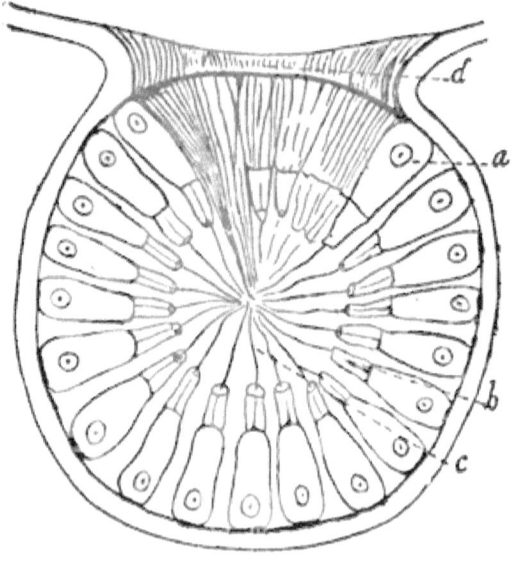

FIG. 2.

try to eat a sponge. The hard parts of a body, supporting and protecting softer ones, are the *skeleton*. We have only the skeleton of our sponges. Which side was fixed to the rock? We are sure it was the smooth, dark side, because it appears to have been cut, and also because some of the sponges have bits of rock caught in the fibres on this side.

Where does the sponge get its food? From the water

it takes in through the tubes. It takes in water through the small tubes that we see, and the tiny ones that we cannot trace carry it all over the sponge. When the sponge has taken from the water the very smallest plants and animals, which are its food, and has given carbonic acid in exchange for oxygen, then the water passes out through the large tubes. But as only the most minute plants and animals can pass through the microscopic tubes without danger of choking them up, a thin, porous skin

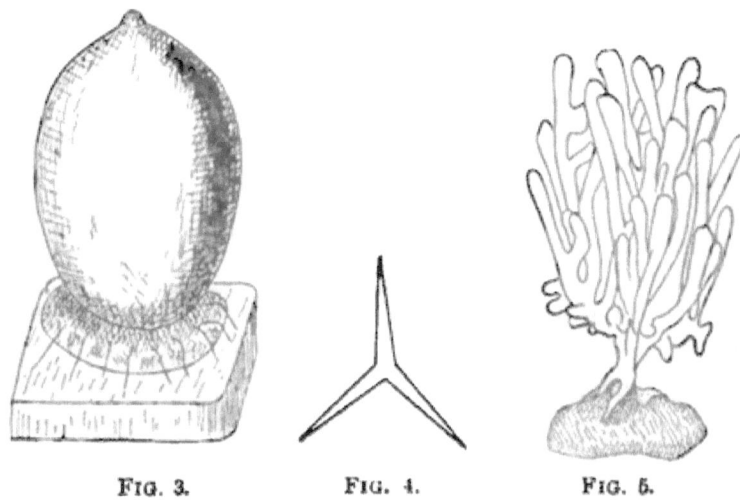

FIG. 3.   FIG. 4.   FIG. 5.

like a delicate sieve covers the whole sponge except the two or three large openings. But why does no water enter at these? Because there is always a current flowing out from them.

In little sacs (Fig 2*) all over the sponge are cells (*a*) bearing each a microscopic whip (*c*), always lashing the water and producing the currents that carry the foul water out through the large tubes as fresh streams come in through the small ones. In these cells, too, the food is digested. Though the outward current keeps the large tubes open, yet if a living sponge is disturbed, it will contract so forcibly as to close even these openings.

---

* Figs. 2 3, and 4 are highly magnified, while Fig. 5 is much reduced from the natural size.

Baby sponges can swim about in the water, but they soon form a sucker at one end, by which they fix themselves (Fig. 3) to rocks, shells, or even the sea fans and other branching corals, and after that they never leave their home unless something tears them off.

Many sponges grow on our New England coast, but are too brittle to be of any use. A little white sponge that grows among shells in the mud just below low-water mark, consists of small branching tubes about an inch long. It has no fibres in its skeleton, but everywhere in its flesh are little three armed bits of lime called spicules (Fig. 4).

The common finger-sponge (Fig. 5) grows in large masses on rocks and piles. The dark red and soft yellow masses found in salt water, and the white flattened cakes often cast up on the shore and dried hard in the sun, are all sponges, the last named called by the sailors "seamen's biscuit."

In a quantity of oyster shells there will usually be one or two, at least, that have been attacked by the boring sponge, which tunnels them through and through, and finally destroys them by dissolving out all their lime.

THE HYDRA.

A tiny green or light brown jelly-like lump as large as the head of a small pin; a slender stem, perhaps one fourth of an inch long, with several transparent threads waving from its tip; or a graceful vase with a few blunt projections from the top; any or all of these clinging to water-plants or any other support in fresh-water ponds or quiet streams, may be the hydra we are seeking. We tell the children we are going to look for a new animal, and invite some of the older ones to join us in a walk, thus enlisting their interest in the hydra in advance. It is a wise precaution to fill our glass jars at several different ponds, rather than to fill several jars from one pond, because the fact that hydras have been found in a given place one year seems to be no guarantee whatever that they will be found there the next year. After the water has settled, the hydras will expand and many of them will collect on the sides of the jar. A dozen watch-crystals filled with pond-water, each with a bit of duckweed or some other green water-plant, make excellent ponds for as many hydras, in which children can examine them with magnifiers, watch them eat, and observe the different shapes they assume. Pupils that are old enough will enjoy making a series of drawings showing these different forms.

Four questions may be put upon the blackboard, *one at a time*, for the children to answer from their own observations: What is a hydra? Where does it live? What can it do? How do new hydras grow?

After a few days the answers to these questions will bring out many of the following facts, in addition to those already mentioned.

The hydra is a green or brown tube, attached by its lower end to some support, and sending out several tentacles near its upper end. Fig. 1 shows one magnified many times, hanging mouth downward, from a bit of wood. The upper end of the tube beyond the tentacles is called the proboscis. At the end of the proboscis is the mouth, an opening leading into the central hollow or stomach.

The tentacles are hollow, like so many glove fingers pushing out around the mouth. They are the hydra's fishing rods, bearing numbers of little pockets, — the thread-cells—on their sides, in which the fishing lines are coiled up. Each line, instead of a hook at the end of it, has three poisoned darts just where it issues from the pocket.

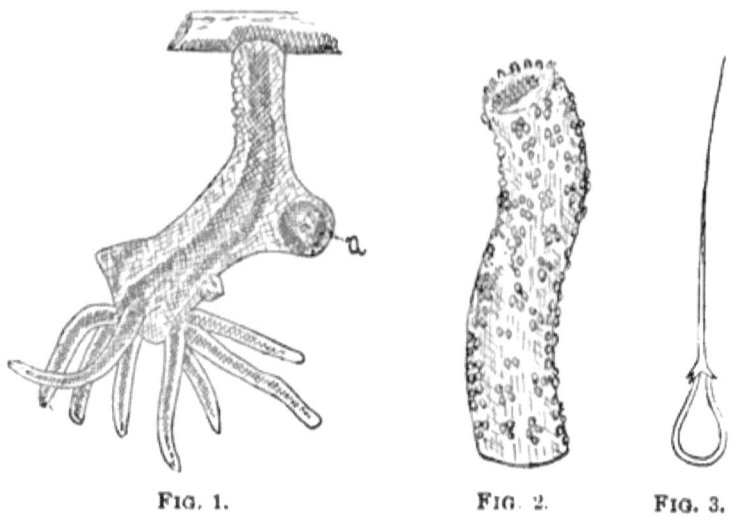

FIG. 1.   FIG. 2.   FIG. 3.

Fig. 2 represents a portion of a tentacle highly magnified, with the thread-cells in clusters on its surface. Fig. 3 is a single cell after it has burst and the thread uncoiled.

When a tentacle touches a tiny worm or crustacean, the pockets burst, and the lines entangle the prey in their coils, while the poisoned darts quickly paralyze it. If some creature too large to be paralyzed is caught by the lines, then ensues a grand "tug of war" between that and the hydra.

 I once watched such a struggle between a hydra and the larva of an insect, which lasted an hour and three quarters. Even then the result was doubtful, but unfortunately the dish containing the combatauts had to be moved, and the stirring of the water shook the exhausted animals free from each other.

The hydra is extremely sensitive, and contracts at once if touched. The variety of shapes it can assume, especially when digesting its food, is very wonderful.

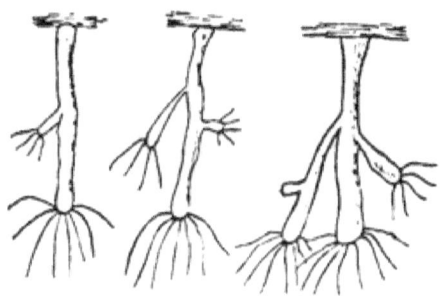

FIG. 4-6.

There are buds on some of the hydras, which at first look like knobs, then grow larger, form tentacles (Fig. 4 – 6), and gradually pinch themselves off from the parent, and set up for themselves. In the autumn eggs are produced (Fig. 1 a), which live through the winter.

FIG. 7-11.

In 1744, Trembley, a watchmaker of Geneva, performed a remarkable series of experiments upon hydras. He found that they can move about by turning somersaults (Figs. 7-11); that if cut in small slices, each slice becomes a complete hydra; that if slit in various ways, a whole colony may be produced from one (Figs. 12-13); and when one is turned inside out, it goes on eating, and appears to enjoy life quite as much as before.

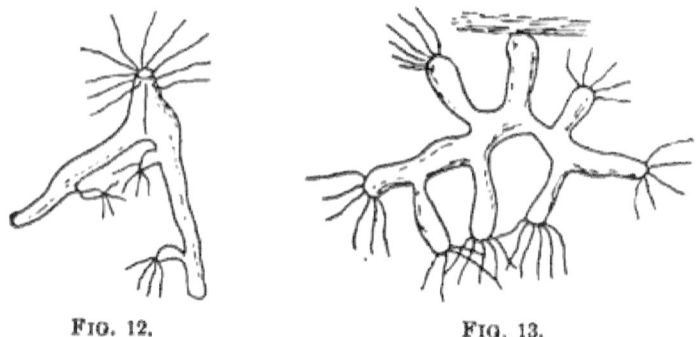

FIG. 12.   FIG. 13.

In tide-pools and on the seaweed along our coast we find graceful, delicate, flower-like clusters, often mistaken for sea-mosses. These are hydroids, or hydra-like animals, whose buds remain connected and form colonies.

THE SEA-ANEMONE.

With living anemones this lesson can be made intensely interesting; without them it should not be given to children. Those who are not too far from Boston can send jars to the superintendent of Essex Bridge, Salem, Mass., who will fill them at a reasonable price. Young anemones can be brought as far as Boston, at least, simply packed in wet seaweed. Nowhere else on our northern coast can such a number or variety of anemones be seen as at Beverly Bridge. The light pink or salmon colored ones show the structure best, for

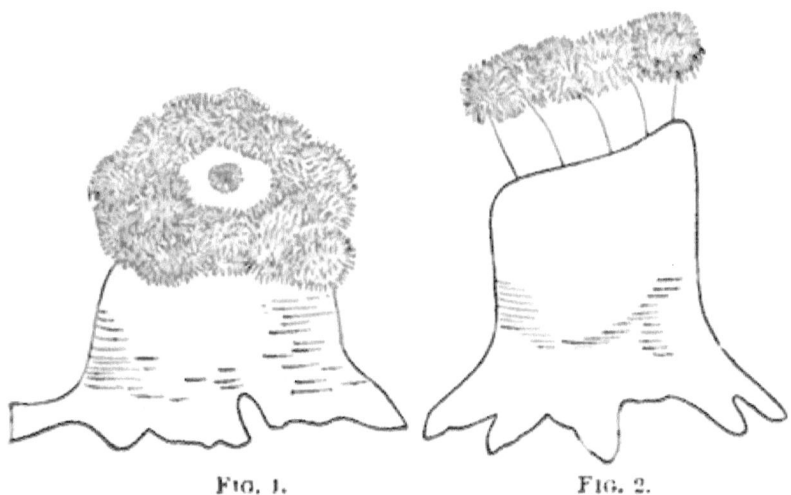

FIG. 1.   FIG. 2.

when fully expanded they reveal the partitions of the body-cavity beautifully through their nearly transparent walls. Fig. 1 represents one fully expanded; Fig. 2, only partially. Very young anemones, not more than an inch long, will show the connection with the hydra, both on account of their small size and of their having fewer tentacles than the full-grown ones.

It has been my experience that large ones do not thrive in confinement, but they can be kept for some days if but one is placed in each jar, and care is taken to keep them cool. Candy jars are best, because the wide mouth allows the air free access to the water. If enough sea water cannot be had to change that in the jars every day, it may be aerated by pouring it back and forth a number of times in a current of air. Young anemones may sometimes be induced to eat meat, drawing it into the mouth with their tentacles, and will generally take the little crab found in the gills of oysters, which they consider an especial dainty, but I have never been able to tempt the older ones with anything. They will not expand well unless kept in a cool, shady place, that shall at least remind them of the tide-pools where they hide under the shadow of the rocks and seaweed. If we try to handle them, as in taking them off the rocks, they contract into little solid lumps (Fig. 3), with apparently neither mouth nor tentacles.

The children should examine them for several days, then bring the results of their observations to the class, as was done with the hydra.

The body is hollow and cylinder-shaped, but very much larger and broader than that

Fig. 3.

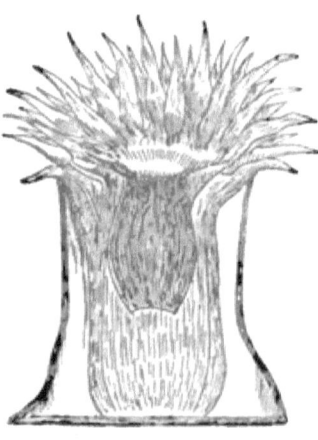

Fig. 4.

of the hydra. By the lower end it attaches itself to some object, and the upper end is a broad disk with the mouth in its center. The mouth seems to be an opening produced by folding the skin inward. Around the mouth are many rows of tentacles, which are finger-like projections from the body.

In the center is the stomach. Many partitions extend inward from the body-wall, some of which join the stomach and hold it in place, others reach only a part of the way to the stomach. They are shown like the spokes of a wheel in the cross-section given in Fig. 5, with the eggs attached to them. In very transparent anemones it can be seen that the tentacles project over the spaces between the partitions. The stomach is not simply a cavity hollowed out in the body, as in the hydra, but is another hollow bag hanging down inside the outer one.

When very young the sea-anemone is like the hydra, but as it grows, the upper end of the body-tube folds inward till it hangs down inside as an open sac, about half as long as the body. To illustrate this, take a glove-finger and cut off the end to represent a hydra, then turn in the end for some distance, and the part hanging down inside will represent the stomach of the anemone. The proboscis of the hydra and the stomach of the sea-anemone are therefore precisely similar in their origin, though different in their use; that is, they are homologous. While digestion is going on, the lower end of the stomach is closed by muscles, and refuse matter is afterward ejected through the mouth.

FIG. 5.

The tentacles are covered with lasso-cells, or thread-cells, similar to those of the hydra. The white threads thown out from tiny loop-holes in its sides when an anemone is disturbed, also bear myriads of these little weapons.

Sea-anemones are often produced from buds, which form around the base of the old ones. If one is torn in scraping it from the rocks, the portion left behind will become a perfect animal.

CORALS.

Lesson I.

A single large specimen of Galaxea (Fig. 1) or some other coral with large tubes, will furnish every child in a class with a tube for study, while the teacher should have a piece consisting of three or four tubes, and, if possible, one or two smaller ones just budding out. Though Galaxea is best, still no teacher need omit the lesson, if she can obtain pieces of the common madrepore or finger-coral (Fig. 2). But if this is used, each child should have the end of a branch showing the large polyp at the tip, and a group of little ones around it. A living sea-anemone in the schoolroom will be a great help. Blackboard drawings of budding hydras should also be kept for the lesson.

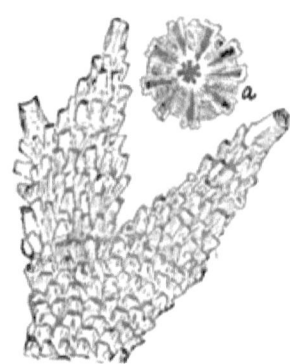

Fig. 1.   Fig. 2.

The children have become familiar with the idea of the skeleton in the sponge, so they at once see that coral is only the skeleton of the coral animal, and that each tube is made by one animal. They quickly make the following observations:

It is white. It is shaped like a tube. It has lines on the outside. It has little walls on the inside. It is hard like stone.

The teacher tells them that this is a stony coral, with a skeleton made of lime. Then they look carefully at the top and the sides of the skeleton, to see if it will remind them of any animal they have studied, and find it is like the sea-anemone.

Some pieces of Galaxea will show plainly that there are twelve stony partitions that nearly meet in the center of the tube, and twelve more that are shorter, but the specimens are often so broken that it is difficult to tell how many of the partitions are long and how many are short. It is not best to have the children count them unless the teacher knows from personal examination of the tubes that her pupils can readily see how many little walls of each sort there are.

After the question, How are these tubes held together? an examination of the teacher's large specimen shows that a stony, white, spongy substance connects them.

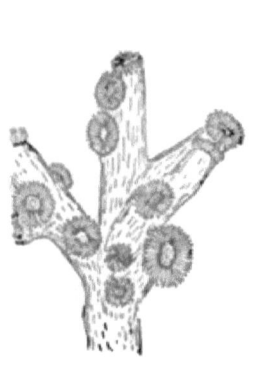
FIG. 3.

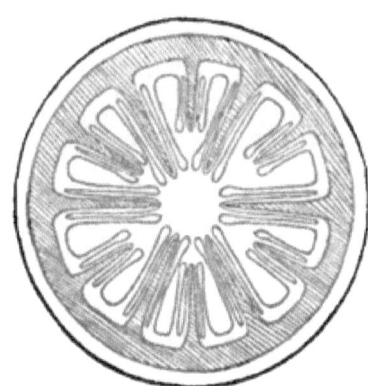
FIG. 4.

Fig. 3 has been put on the blackboard, drawn wholly in red, because it shows only the fleshy parts of the coral. This is not the Galaxea, but it has the same kind of a stony skeleton, and the same arrangement of all the fleshy parts. The children now describe this figure.

This new coral has a fleshy tube. It has a disk at the top of the tube, with the mouth in the center. It has tentacles around the mouth. There are little animals budding from some of the tubes. There is flesh covering the stony skeleton between the tubes.

It is easy now to understand that the spongy filling between the tubes of Galaxea is formed by the layer of flesh that covers it, and connects the animals. A colony of Galaxea is formed by the budding of young animals from this connecting layer, around the base of the old ones.

Fig. 4 is a cross section of the body of a living coral, but does not show the stomach. It represents what we should see if we were to cut off the upper half of the tube and then look down upon what was left. For the blackboard the unshaded parts should be drawn in red, to represent flesh, and the shaded parts in white, for the stony skeleton. The children now tell what they see in this figure:

There's a tube of flesh outside of the tube of stone. There are fleshy partitions and stony ones. The fleshy partitions are in pairs, and the stony ones are not. There are six pairs of long, fleshy partitions, and six pairs of short ones. There are six long stony partitions, and six short ones. There is a tube of flesh inside the stony tube, and the fleshy partitions grow out from that.

The stony partitions are not formed by the fleshy ones, but grow in folds of the fleshy tube, that arise from its base between the fleshy partitions.

Lesson II.

The same specimens are needed as for the last lesson, with finger-coral besides.

Review by asking what the coral has that the sea-anemone has, then what the coral has that the sea-anemone has *not*. This comparison will bring out from some bright child the observation, "Why, the coral is just like a little sea-anemone with a skeleton!" This is exactly what we wish to find out, and the children will see it more clearly if there is still a little colony of very small sea-anemones in the schoolroom.

Fig. 5 is a diagram to be drawn in red and white, showing the upper half of the fleshy tube with the tentacles, and a cross-section through the calcareous tube with its partitions. To make it plainer, only six long and six short partitions are shown, and just as many tentacles, each long tentacle lying directly above a long partition, and each short tentacle above a short one. This figure not only shows the resemblance between the coral animal and the sea-anemone, and the relations of the fleshy parts to the calcareous skeleton, but also the manner in which the fleshy base of the tubes (*b*) extends from one to another, thus binding together all the animals. It is this which forms the spongy filling of lime (*c*) between the tubes, and from which new tubes bud as the colony grows. A vertical section through a large piece of Galaxea will often show that most of the polyps die at the end of each season; but the next season those that have lived spread out their fleshy bases and build a new spongy layer of lime over the tops of the dead tubes below.

FIG 5.

The children will have suspected by this time that the stomach, the mouth, and the poisoned lines on the tentacles of the coral are like those of the sea-anemone, as is really the case.

The class is now ready for the finger-coral, and the teacher may happen to have a piece that resembles a hand with fingers, thus suggesting its name. The children's observations follow:

It is white and stony. It grows in branches. It has little bits of cups on the branches. It has very small tubes. The tube at the end of the branch is much larger than the others. The large tube has partitions on the inside and on the outside too, but the little tubes are only rough. I think that is because the tubes are so small that the partitions can't grow very far. The smallest tubes of all are close to the large tube at the end of the branch. They look as if they had budded from the large tube.

Some of these observations must be drawn out by such questions as will readily occur to any teacher.

Now if a branch of the coral is broken and passed around the class for the scholars to examine the broken ends, while one with an eye for beauty, may see "something like lace with a star in the middle," and another only "a piece of stone with a little wheel in it"; some one will finally discover that "the middle of the branch looks like a cross section of a tube" (Fig. 2a). In this way children can find out for themselves that the large tube at the end of a branch, which has kept on growing year after year, is the one from which all the rest have budded. Great bushes of this coral sometimes grow to the height of sixteen feet, so we know that the parent polyp of each branch must live to a great age.

The stony corals are the reef-builders, and a large part of every coral reef consists of branches of madrepore beaten and pounded by the waves into a mass of coral rock. These lessons on stony corals may well be followed by an imaginary trip to the coral islands for a geography lesson.

Lesson III.

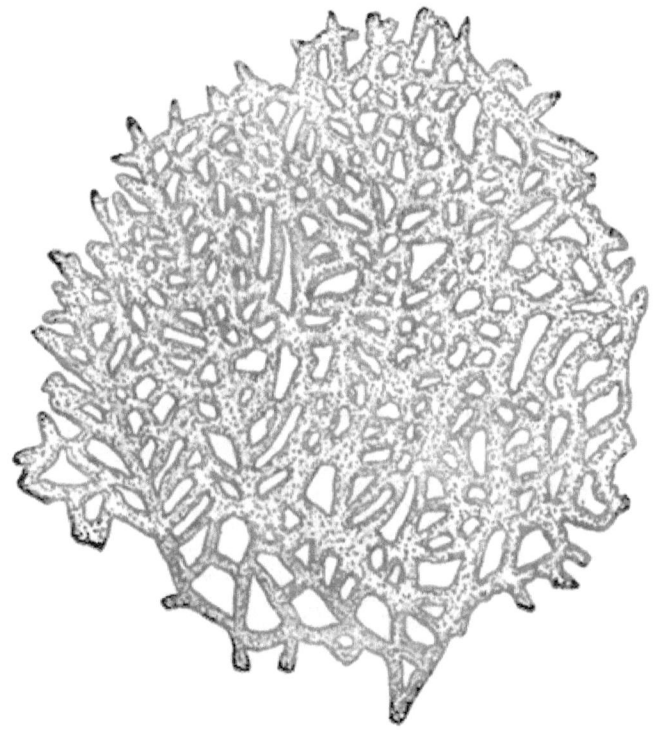

Fig 6.

A whole sea-fan (Fig. 6) for the teacher, and small pieces for the children, with the stony corals before used, are needed for this lesson. These should be supplemented by a small piece of the precious red coral, and by a blackboard drawing of Fig. 7 in colors. In this red-coral the branches themselves, as well as the flesh covering them, are red, while the separate polyps are white.

The name fan-coral, or sea-fan is quickly suggested as the teacher's large specimen is held up before the class, after which pupils' observations are in order:

My piece of sea-fan is yellow. Mine has dark edges. Mine is made of a great many little branches, that join together in a network. I can break off something like a yellow crust from the edges of mine, and there's a little dark brown wire left. I can see lots of little dots all over mine. They look like pin-holes.

Fig. 7

All are now interested to learn that the "yellow crust" is the flesh that connected all the animals of the colony, and the "brown wire" is the skeleton. The flesh contains so many bits of lime that it becomes hard when it dries, and so remains on the stems. The skeleton is horny. In order to understand what the "pin-holes" are we must turn to the diagram of the red coral (Fig 7), which the children describe:

There is red flesh between the coral animals. The coral animals are white. They have only eight tentacles. The tentacles are fringed. One of the coral animals has drawn itself back into the red flesh, and only its tentacles show. Two other coral animals have drawn themselves all back into the red flesh, and there's only a little bit of white in a round, red place to show where they were.

It is now easy to see that the animals which make the Galaxea and the finger-coral could not hide so nicely in the flesh that covers the branch because the stony tubes would be in the way, and to draw the conclusion that the red coral has no separate tubes. Neither does the fan-coral animal make tubes.

Skillful questions now lead the pupils to see that if the red-coral animals die, and the flesh dries, a red stem will be left with dried flesh on it, and in the flesh little holes that show where the animals were. So they know that the "pin-holes" in the yellow flesh of the sea-fan are the places where the living animals were.

The sea-fan and the red-coral, as well as the others studied, increase by budding, but in all these, each separate colony starts from a single egg.

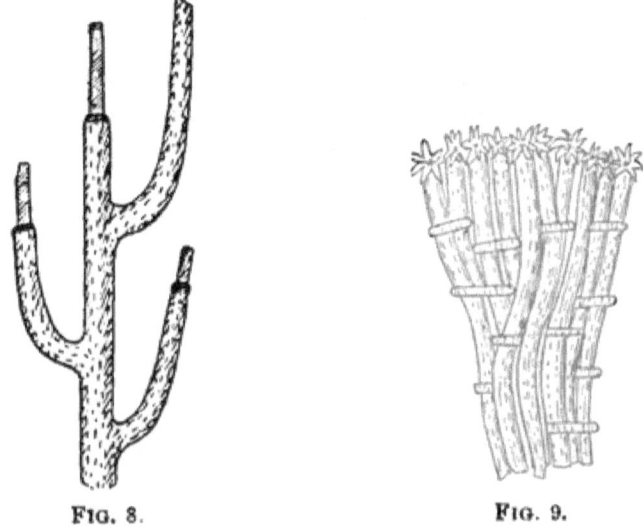

FIG. 8.   FIG. 9.

The lesson ends with a comparison of the stony corals and the sea-fan, which the children afterwards write out.

Fig. 8 represents a coral like the sea-fan in structure, in which the branches do not interlace. Fig. 9 is the organ-pipe coral with its green polyps expanded above the red tubes.

THE STAR-FISH.

Lesson I.

Specimens: A dried star-fish and a single dried ray for each child, and a few large star-fishes in alcohol. The single rays should be cut open down the back and the contents removed, leaving only the sacs connecting with the tube-feet. The best way to prepare star-fishes dry is to put them into fresh water until their bodies become fully rounded, then to soak in alcohol for au hour or two to harden the tissues, and finally to dry in the sun or a moderately hot oven. If alcohol is too expensive, they may be put into boiling water for a few minutes before drying. But the rays will drop off if they are left in the hot water too long. In this lesson, and the two following, parts are described as they appear on *dried specimens*.

Fig. 1.

Our new friend is called a star-fish, from its shape like a star with five points, which we call rays or arms. Its home is in the sea, on the piles of wharves, among the rocks, or on oyster and mussel beds. In summer it is often found above low-water mark in tide pools, but in winter it takes refuge in deeper water. When dried it gives us no idea of the rich colors,—red, bluish, green, or brown,—with which it beautifies the sea-bottom. Unlike most of the animals so far studied, it can move slowly from place to place.

The children at once find the mouth, but we wish to study the back first, "the side that is rough all over." (Fig. 1.) Holding this uppermost we find the sieve, a round, coral-like spot that is red or orange when the star-fish is alive, and used to filter the water that passes in through it. The central part of the star-fish is the disk. The sieve is on one side of the disk, near the angle where two rays meet. If now a line is drawn from the sieve across the disk and through the middle of the opposite ray, there will be the same number of rays on each side of the line, that is, the star-fish will be divided in halves.

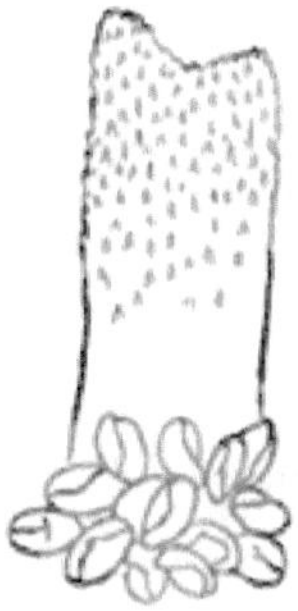

This gives us a hint of the bilateral symmetry seen more perfectly in the higher forms.

The back of the star-fish is covered with knobs,—"prickers," the children may say,—called spines. They are short and rounded at the tip, and we find by trying them on alcoholic specimens, that they do not move. Between and around the spines are little things looking like tiny grains of meal, which are two-pronged forks, or pedicellariæ (Fig. 2), always opening and shutting. We do not know their use, unless it is to keep dirt from clinging to the star-fish. On alcoholic specimens these will

FIG. 2.

show beautifully in little circles around the spines.

Covering the star-fish we see the brown skin, and looking on the inside of the back of the separate rays, we find the beams of the skeleton, like little bones imbedded in the flesh, making an irregular net-work.

Every one who has seen a living star-fish, has noticed the difference between its rounded outline in the water and its flattened appearance when thrown up on the beach. This is because the skin of the back is pushed out into numerous tubes like tiny glove-fingers (Fig. 3, d), so thin and delicate that they fill with water and again allow it to ooze out of them when exposed to the air. These tubes, that can scarcely be seen by the naked eye, are really also a rudimentary sort of gills, for through their thin walls oxygen passes from the water into the body of the star-fish.

FIG. 3.

LESSON II.

We now study the under or mouth side of the star-fish. Some of the specimens will show the mouth as a large circular opening with a membrane surrounding it; others will have a brown mass, the dried stomach, filling the opening or protruding from it; and still others may have it nearly hidden by ten long spines, two from each ray, meeting over it like so many teeth.

The mouth with the long spines around it, the stomach usually seen just inside it, and the brown suckers filling the grooves in the rays (Fig. 4) first attract our attention. The stomach can be protruded by means of muscles attached to it. This is because our friend feeds on shell-fish, working great havoc on the oyster and mussel beds. It clasps an oyster with its rays, then turns out its stomach, and proceeds to digest its victim at its leisure. A star-fish will clean a shell in this way more perfectly than it can be done by hand.

The four rows of suckers in each ray are on the ends of the tube-feet. In alcoholic specimens these completely fill the grooves, and in life they even extend beyond. The star-fish moves about very slowly. stretching out one ray as far as possible in front, planting a few suckers at a time, drawing the body up to them, then lifting them and taking a fresh start. The tube-feet at the tip of each ray are extended in front as feelers.

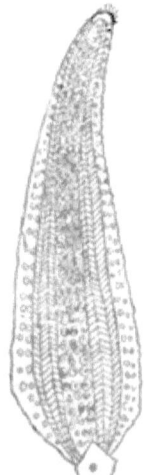

FIG. 4

The sieve on the back connects by a tube with lime in its walls, hence called the stone canal, with a circular canal around the mouth, from which a tube extends down each ray. From these radial tubes branches lead to each one of the small, muscular sacs (Fig. 3, *g*) seen on the inside of the rays to be connected with the tube-feet. These sacs and some larger ones opening into the circular canal, act as reservoirs for the water that enters at the sieve and force it down into the tube-feet when the star-fish moves. One can fully understand and appreciate this water-system of the star-fish only by seeing it in a specimen in which the tubes have been injected with coloring matter.

Down the middle of each ray a brown line.—the radial nerve (Fig. 3, *f*),—will be seen on most of the specimens ending at the tip of the ray in an eye. In life the five little red eyes filled the tiny hollows at the end of the rays.

The central part of the nervous system, as of the water-system, is a cord around the mouth, occasionally seen on the dried animals. The eyes of the star-fish see light only, as ours do when the lids are closed.

The star-fish has also the sense of smell. After one has been kept without food for several days, it can be led around the tank after a piece of shell-fish held just in front of it with a pair of forceps, precisely like a hungry dog after a bone.

Lesson III.

The hard parts of the back were noted,—the spines, the forks, and the beams of the skeleton,—now those of the mouth side are to be examined.

On the cross-section of each ray (Fig. 1) may be seen two rows of narrow plates (Fig. 3, *a*) forming the roof of the groove, with small openings between them through which the tube-feet pass. These are the perforated plates. On each side of these is a single row of irregularly shaped, somewhat thickened plates (Fig. 3, *b*), called in distinction from the others, unperforated plates. On the alcoholic specimens we observe that the slender spines borne on these plates are movable, the only movable ones, in fact, on the body of the star-fish.

Inside the rays are the brown masses of the liver and the grape-like clusters of the ovaries. The eggs pass out at minute openings, difficult to find, in the angles of the rays.

Not only is the shape of the body radiate, but all the organs show a radiate arrangement. From the central water-system run five radial water-tubes; from the stomach a lobe to each ray; liver-lobes are found in each; the oral nerve-cord sends out five branches, and each ray ends in an eye. The attention of the class may be drawn to as many of these points as they have observed, in order that they may get the idea of the symmetrical arrangement of parts around the common center.

Some of the specimens will show the convenient power that the star-fish has of replacing lost rays (Fig. 5). He seems not to mind the loss of two, three, or even four, at a time, and grows them again with wonderful rapidity More than this, if he is torn in pieces, and the parts thrown into the sea, each ray will become a star-fish.

The Star-Fish

This star fish with the round disk and the snaky arms is the brittle-star (Fig. 6). Though found in abundance on our coast, he hides himself away under rocks and seaweeds, so that he is rarely seen by most people. "Catch me if you can," he seems to say, for not only does he wriggle away at a speed that is the greatest contrast to the snail's pace of the common star-fish, but if we do seize one long ray, he coolly drops it off, and disappears with the other four. Often the only way to secure a perfect one is by dipping it up in a quantity of water with the seaweed on which it lies.

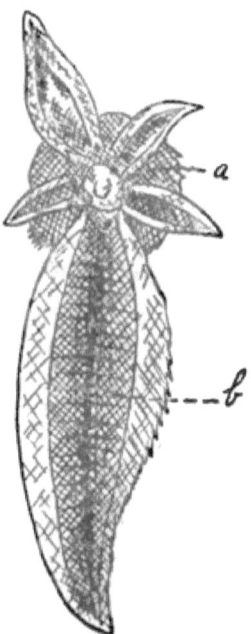

FIG 5..

FIG 6

The rays of the brittle-star are solid, the ambulacral plates,—which correspond to the perforated plates of the star-fish,—being on the inside and covered by a row of extra plates, so that the organs of the body must needs be all in the disk. Though his tube-feet end in points, he does not miss the suckers, but moves about by means of his long, muscular rays. In order to move rapidly he uses two of his opposite arms with a motion like swimming, leaping about two inches at each stroke.

A small, beautifully colored star-fish, also found on the New England coast, has five tapering rays with only two rows of tube-feet on each. This species carries its eggs around the mouth.

The common star-fish south of Long Island Sound is green or brown, the one usually found north of that sound is red or purple.

THE SEA-URCHIN.

Lesson I.

Not the little ones usually seen on the shore after a storm or in the tide-pools, but large ones that have been taken from their hiding places under stones below low-water mark. The best preparations for study are made by drying them with the spines on, and then sawing them in two horizontally. Besides these every shell and piece of a shell without the spines will be of use. Perfectly bleached shells are often cast up on the shore by the waves, but if enough of these cannot be found, the spines may be removed by placing them for a time in a dilute solution of potash, then cleaning them with a tooth-brush. Care must be taken not to leave them in the potash too long, or it will cause the plates to drop apart. It will add greatly to the lesson if the teacher can also have a few large sea-urchins in alcohol, and a Mediterranean Echinus, or sea-egg, either with or without the spines.

The sea-urchin (Fig 1) is also sometimes called the sea-egg. It is a green or purple ball. It is not a perfect ball, but flattened on both sides, more so on the mouth side than on the back. It is bristling all over like a small hedgehog with spines longer than those of the star-fish.

We try the spines on the alcoholic specimens and find them movable. They are held to the shell by tiny muscles. We carefully pull off a large spine, and see the knob upon the shell and the socket in the base of the spine that fits over the knob, thus making a ball-and-socket joint.

We know the mouth on the under side (Fig. 2) by its little star made of the five white teeth. Though the teeth appear small when seen from the outside, on examining them from the inside we find that with the jaws and muscles attached to them they form a powerful apparatus called the lantern. Inside the shells from which the lantern has been removed, can be seen five projections to which some of these muscles were attached. With its teeth, which are constantly growing at the root as they are worn away at the tip, the sea-urchin scrapes seaweeds off the rocks, as it walks about mouth downward, and it also feeds upon dead fish.

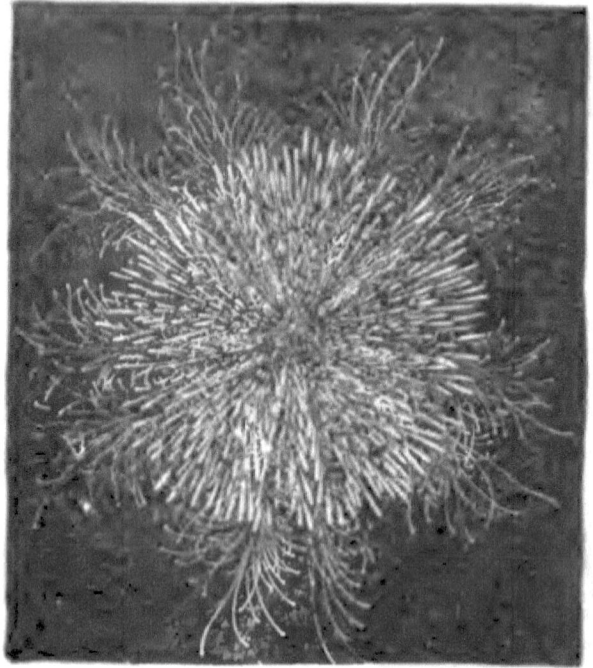

FIG. 1.

The sea-urchin walks with tube-feet, as the star-fish does, but its suckers are smaller and more nearly the color of the spines. If we have only dried specimens, in order to see the tube-feet we must look steadily and patiently on the under side of the shell. Children will see, after a little thought, that they are largest on this side because they are used most in walking. But how can the sea-urchin walk on the suckers, while the spines reach out so far beyond them? To understand this and to know how beautiful the little creature really is, we ought to have a living one in a glass dish or jar of sea water, and watch it press out its tube-feet till they are waving on every side, far beyond the spines, like threads of spun

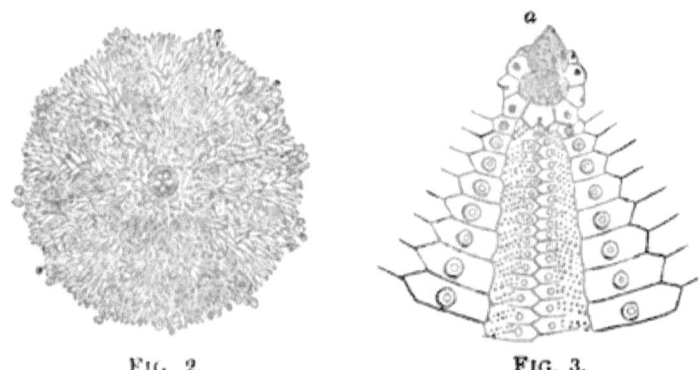

FIG. 2.   FIG. 3.

glass. It stretches them out to their fullest extent, fastens the suckers, then pulls the body up to them. We touch it, and instantly every one is drawn safely in behind the barrier of sentinel spines.

Since the tube-feet of the sea-urchin are forced out in the same way as those of the star-fish, we see the need of a sieve to filter the water that does this work. We find it, a five-sided, spongy body (Fig. 3, a), at one side of the little disk in the center of the back.

Lesson II.

In the last lesson we have examined the spines, have found the mouth with its lantern, the tube-feet and the sieve, and have learned how the tube-feet can be extended beyond the spines. We begin with a review of these points.

The spines protect the sea-urchin against its enemies, and are sometimes used in walking. If one is taken out of the water and put on a table, it will try to walk on its spines.

Some species habitually use the spines in this way. A pentagonal sea-urchin that lives on coral reefs, covers itself with bits of seaweed and pebbles, holding them on with its tube-feet, so that but little of its body is exposed, while walking on its spines. Our common sea-urchin often hides under a covering of sand and gravel.

On alcoholic specimens, and on the under side of large, well-preserved dried ones, children will see that the tube-feet are arranged in five double rows, and later in the lesson will connect this fact with the arrangement of the plates of the shell. The forks can be best seen on the disk of tough skin around the mouth, where they form a circle. They are larger than those of the star-fish, three-pronged, and mounted on handles (Fig. 4). They are scattered everywhere among the spines, but largest around the mouth. It has been found by experiment that they will grasp a frond of seaweed waving lightly over them in the water, and hold it like so many tiny forceps till the suckers have had time to fix themselves upon it.

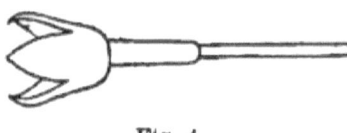
Fig. 4

The remainder of this lesson is the most difficult part of the work on the sea-urchin, and should be conducted by the teacher with the greatest patience and care. Though "Make haste slowly" should always be the motto in science lessons, any attempt at haste will be especially fatal here.

Taking the bare shells we first see that they are composed of many parts called plates. As we hold them up to the light and look inside, we discover a great many small openings in them, like fine pin-holes. There are ten rows of these openings, or five double rows, and sharp eyes will see that there are also five double rows of small plates through which these holes pass. These are the perforated plates, and the holes are for the tube-feet, as in the star-fish. On each side of a double row of small plates is a double row of large plates without openings. These we call the unperforated plates, and we find five double rows of these also. We remember that there are ten rows of perforated plates in the star-fish, also, as well as ten rows of unperforated plates.

We now find an opening in the sieve, and we observe that the sieve is at the end of two rows of large plates. From the inside of the shell four more openings can be seen at the end of the other double rows of large plates. These are the openings through which the eggs pass out into the water. The egg-openings are in little plates shaped somewhat like the sieve, and really five in number since the sieve is on one of them.

Five more tiny holes alternate with the egg-openings, and stand at the end of every two rows of perforated plates. These are eye-openings, and are in very small plates called eye-plates. The young sea-urchin has five eyes in these places, but when fully grown he has only these orifices, through each of which a tube-foot passes out. Inside the circle of eye-plates and egg-plates is a little disk of tough skin containing minute plates.

Both egg-plates and eye-plates are difficult to make out clearly, except on large specimens of the common sea-egg, but both are shown beautifully on the large Mediterranean sea-urchin.

Each child should now draw such a section of the shell as is represented in Fig. 3, consisting of two rows of perforated plates with one row of large plates on each side of them, and the central disk surrounded by its circle of plates.

Lesson III.

When we go to the seashore, where shall we look for star-fishes?

In tide-pools. On the rocks under water. On oyster and mussel beds.

Where shall we find sea-urchins?

In tide-pools. On rocks under water. Sometimes hidden under stones and gravel in the water.

What did we find on the star-fish?

Short spines. A sieve. Tube-feet. A mouth. Five eyes. Five nerves. Tiny specks of forks.

What have we found on the sea-urchin?

Long spines that will move. Tube-feet. A sieve. A lantern. Five teeth. Lots of forks with handles. Five holes where the sea-urchin had eyes when it was young. Rows of small plates with holes in them. Rows of large plates.

How many rows of each?

Ten rows of each.

What plates did we find on the star-fish?

Plates that made a network on the back. Two rows of perforated plates on each arm. One row of unperforated plates on each side of the perforated ones.

What is the use of the holes between the perforated plates of the star-fish?

The tube-feet pass through them.

How are the rows of plates arranged in the sea-urchin?

First there are two rows of large plates, then two rows of small plates.

How many rows in all?

Ten rows of each kind.

Where are the eye-openings?

At the end of the rows of small plates. At the end of the perforated plates. At the end of the plates that the tube-feet come out through.

Where are the eyes of the star-fish?

At the point of each ray.

Are they at the end of any rows of plates?

Yes, at the end of the rows of perforated plates.

And where are they in the sea-urchin?

At the end of the rows of perforated plates, too.

Where are the egg-openings?

At the end of the rows of large plates.

But we have not found any nerves, some child suggests. No; for the sea-urchin has them safely tucked away on the inside of its shell.

Fig. 5 is a diagram representing a cut through a sea-urchin; *s* is the sieve, or madreporic body, and *c* the tube leading from it to the ring around the mouth, from which branch the five radial water-tubes (*f*). Each of these radial water-tubes connects with the sacs (*g*) of the tube-feet and ends in a single tube-foot (*x*), which passes out through an eye-opening. The black line outside of *f* is a radial nerve, also continuous with a cord around the mouth. The cut ends of the intestine are seen at *k*: the ovaries at *l*; the branchial tufts at *m*; at *h*, the teeth; and at *i*, some of the muscles that move the jaws; *b* is a spine; *e*, a plate of the shell; and *n*, one of the forks, or pedicellariæ.

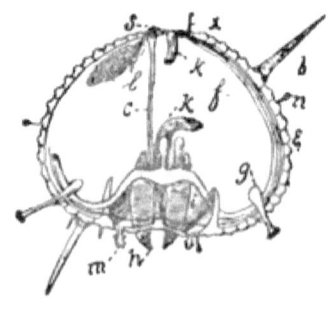

FIG. 5.

If any of the children have sand-dollars (Fig. 6) or sand-cakes, as they are often called, they will be delighted to discover that they are simply flattened sea-urchins. Perfect ones are very pretty, with their mouse-colored covering of tiny spines, and the bare shells show the different rows of plates and the mouth on the under side. The tube-feet project from perforated plates, and form the beautiful star on the back by expanding into gill-like appendages. This sea-urchin lives partly buried in the wet sand of our shores and of the shallow waters.

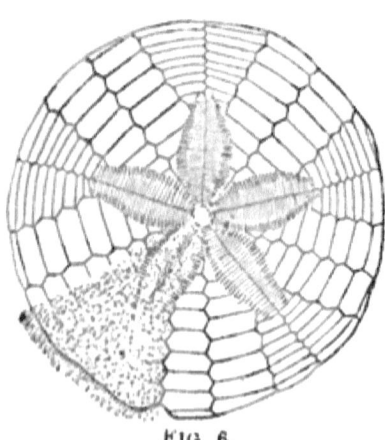

FIG. 6.

THE CLAM.

Lesson I.

The common soft-shelled clam of the New England coast (Fig. 1) is the one chosen for these lessons, but the fresh-water clam found everywhere inland can be used equally well. Large ones should be obtained, and if kept alive in the schoolroom in a jar of sea-water for a few days before the lesson, the children's observations will form an excellent preparation for class study. The long, dark siphon, often incorrectly called the head, will be extended with the two fringed openings plainly showing, but will be at once drawn in if the shell is touched. If some very finely powdered in-

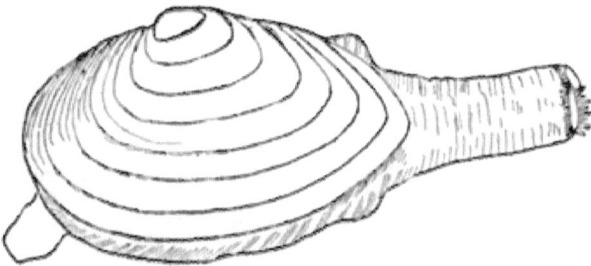

Fig. 1.

digo is dropped into the water, the particles will enter at the lower opening and pass out at the upper one. It will be seen that the valves of the shell are connected by a dark skin that passes from one to the other, and is unbroken except by an opening near the broad end of the shell, where the foot is protruded. If a living fresh-water clam is kept in a jar with two or three inches of sand in the bottom, it will assume its natural position, with its foot extended and its body partly buried in the sand (Fig. 2). With this form particles of indigo can be even more plainly seen passing in and out of the large siphonal openings.

If the clams are killed by putting them into warm water the day before the lesson, the soft parts can easily be removed without breaking the ligament at the hinge, while the two valves are still held together. The shells thus prepared can be put into the hands of young children, who can afterwards see the principal soft parts, such as the mantle, the gills, and the foot, from the teacher's specimen.

The clam shell consists of two valves, which are convex on one long edge and nearly straight on the other, and which meet at a "rounded point," called the beak. It is broad at one end and narrower at the other, but both valves are alike in size and shape. To find the right and left valves we hold the shell with the beak uppermost and the narrow end pointing toward us, when the right valve will be on our own right side, and the left valve on our left.

The surface of the valves is not smooth, but roughened by many curved lines. Beginning at the beak and tracing all the lines between this point and the convex edge of the shell, we find that they all start at the upper side of a valve and pass around to the upper side again. Children will quickly see that the baby clam could have had but few of these lines, and that more have been added as it grew, hence these are lines of growth. Now they will be interested in tracing these back from the outer edge to the beak, and seeing that this is the shell of the baby clam and the oldest part of the whole shell. A brown, horny skin covers the lower part of the valves, but is worn away near the beak.

A few of the thickest shells have been roasted on a bed of glowing coals, and pieces of them are now distributed to the class.

The roasted shells are white and crumble easily. The lines of growth are the edges of layers of lime, which can now be peeled off. The shells are not so heavy as they were before.

Since the shells weigh less after the roasting, it is evident that some part of them has been burned away. To discover what was lost, we suspend one or two shells in a dilute solution of acetic, or hydrochloric acid, which will remove the lime.

The valves are held together by a hinge at the beak. On the inside of the hinge is a brown, horny substance, the ligament. By slowly opening and closing the shell several times, we discover that the ligament (Fig. 3, *l*) is

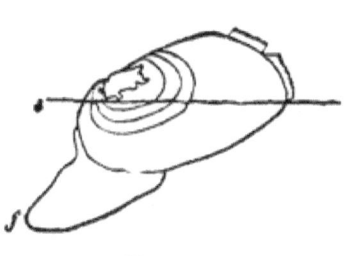
FIG. 2.

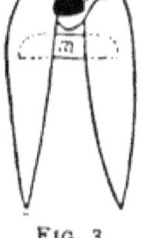
FIG. 3.

compressed when the shell is closed (Fig. 3), but, being elastic, it forces the valves apart when the pressure is removed. On the left valve is a small shell-like tooth, and a corresponding socket on the right one.

The class now draw the outside of one valve, writing on the drawing the names of all the parts they have seen.

Fig. 3, section through a clam-shell; *l*, the ligament; *m*, one of the muscles that draw the valves together.

LESSON II.

In addition to the shells already studied, children who are old enough may have freshly killed clams in which only the adductor muscles have been cut. Specimens put into warm water the day before the lesson, will die with siphon extended. The teacher needs also a living clam.

We first examine the shells that were left in dilute acid:

These shells are soft. They will bend easily. There is no lime in them, but they seem to be made of flesh. Clam shells are made of layers of flesh and layers of lime.

We remove the left valve now and examine the inside:

It is whiter and smoother than the outside. It has a line around it near the edge. It has a broad mark near the narrow end and another near the broad end of the valve (Fig. 4). It has a deep curve between the line and one broad mark.

One child takes the living clam and finds he cannot open its valves. The teacher then holds up a dead clam in which the muscles have not been cut, and after carefully pushing back the fleshy bag that adheres closely to the edges of the shell, she severs both muscles. The children see that when they are cut, the valves come apart everywhere, except at the beak. One child now lays the valve back over the body of the clam, and finds that the two broad marks on the inside (Fig. 1, *aa* and *pa*) exactly cover two firm, white organs, which are the muscles that held the valves together. In this

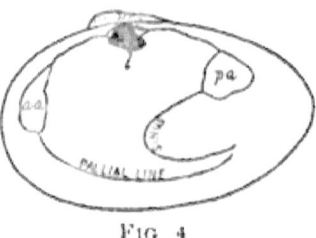
Fig. 4

way see that the marks are the impressions of the muscles, and were made by them.

The same process shows us that the line near the edge was made by the thick, muscular part of the fleshy bag, or mantle, that covers the clam, while the deep curve, or sinus (Fig. 4), is the impression of the siphon muscles.

We have noticed that the dark skin covering the margins of the shell, seems to be connected with the mantle, and, indeed, it is formed by the mantle border. This same border of the clam's cloak is a hard worker, constantly laying down new deposits of lime around the edges of the shell. The mantle is closed except at the siphon and the foot openings.

After finding the little furrow in the lower edge of the mantle and cutting through it from the siphon to the forward end of the shell, we lay the mantle back, as represented in Figure 5. Just under the mantle lie two of the gills (Fig. 5, a), two delicate, fluted ruffles, and on the opposite side of the body are two more. At the forward end of the body are two pairs of delicate flaps, the palpi (Fig. 5, p), and between these is the mouth (m). The foot (f) is a curious little lump of muscle, easily found, because there is but one. Now we see how the clam can burrow in the mud, and why he needs such a long siphon to reach up to the water, while he is snugly hidden in his hole.

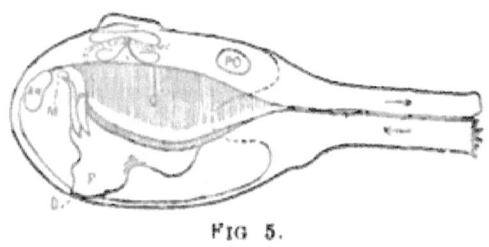

FIG 5.

So much the youngest children can see from the teacher's specimen. They should now draw the inside of one valve, and mark on it the position of the gills, the siphon, the foot, and the mouth, as well as the impressions on the shell. If there is time, older children, each with a clam in a small dish of water, can go further.

A probe passed into the lower opening of the siphon, enters the body cavity just below the gills. Water coming in at this tube is strained by the waving hairs on the gills, then passes through the gills, giving up its oxygen to the blood on the way, and out by the upper tube of the siphon. At the same time by the motion of the hairs, or cilia, tiny plants and animals in the water are gathered into threads and swept down to a channel on the lower edge of the gills, along which they move forward to the mouth. By lifting the palpi a probe can be passed into the mouth.

To see the heart, cut away the mantle from the top of the gills, being very careful not to cut anything else. In a little cavity just behind the beak lies the heart, with the dark tube of the intestine passing through the ventricle. The somewhat fan-shaped auricles, one on each side of the ventricle, are not easily found except by one skilled in dissection.

To be seen distinctly, all these fleshy parts must be floated out under water.

Lesson III.

The last two lessons should be thoroughly reviewed by the aid of the shells and a blackboard sketch on which the organs can be located as they are described by the children.

It will add interest to the talk about the burrowing habits of the clam, if the teacher can show a razor-shell (Fig. 6), and explain that this kind of clam burrows so fast with its powerful foot that one can head it off only by a sudden oblique cut with the spade.

FIG. 6.

South of New York the round clam, or quahog (Fig. 7), is the common one in the market, and will be the most convenient type for these lessons. On sandy shores this burrows but little below the surface, often even crawling about with its shell partly exposed.

FIG 7.

This habit is the explanation of the short siphon tubes. It is taken in muddy creeks by long tongs or rakes.

The large, rounded shell of this clam, with its prominent beak, has the hinge ligament on the outside. Children will see, by carefully opening and shutting it, that the ligament is stretched when the shell is closed, and so opens the shell when it contracts to its natural size.

The foot and mantle edges are white, the siphon tubes yellowish or brownish orange, mottled toward the end with dark brown or opaque white, and are separated a little at the end. The mantle lobes are separate and ruffled at the edges. The clam easily burrows when necessary, by means of its large foot with a broad, thin edge, which can be protruded from any part of the lower side through the large opening in the mantle.

In many parts of the country the fresh-water clam will be the one most easily obtained. The manner of keeping it in the schoolroom has already been described.

A horny brown skin sometimes covers the whole shell of the fresh-water clam, but is usually worn off near the beak by the action of acids in the water.

Fig. 2 shows the position of one of these clams in crawling, the line at *s* representing the bottom of a lake or river, above which is water.

The beak is far forward, and the long hinge ligament, being on the outside, acts as in the quahog, like a spring to open the shell. On the inside of the shell are several

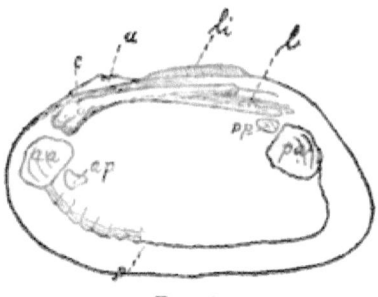

FIG. 8

hinge-teeth, if the clam is a Unio, generally two wedge-shaped teeth on the left valve and one on the right, while back of these the right valve bears one long lateral tooth and the left bears two.

In Fig. 8 two of the wedge-shaped, or cardinal teeth are shown on a *right* valve at c, and the lateral tooth at *l*.

The pearly lining of this shell is the same as that which forms the mother-of-pearl of the pearl oyster. When this deposit is increased by a particle of sand or some small object that has worked its way in between the mantle and the shell, an imperfect pearl is formed. Since the clam cannot expel an intruder, he takes this way of covering it up and thus freeing himself from the irritation it caused.

Near the impressions of the adductor muscles (Fig. 8, *aa* and *pa*) are two smaller prints (*ap* and *pp*) made by the muscles which move the foot. The pallial line (*p*) runs from the anterior to the posterior adductor, but without any sinus. This lack is explained when we see that this clam, the back part of whose body is never buried in the sand, has no long, muscular siphon, because it needs none. The mantle, open everywhere else, is simply united enough at the siphon to make two very short tubes, which are always in the water.

The parts of the body have the same relative position as in the salt-water clam, but the palpi, instead of hanging freely, are simple folds.

THE OYSTER.

Lesson I.

Our oysters, which are the largest ones to be had, will need at least twenty-four hours in warm water to kill them, even if a hole large enough to admit the water has been picked in the edge of each one. They are sponged to remove the dirt, without taking off the brown skin. Roasted and decalcified shells are on hand, too, to show the layers of lime and flesh. Before the lesson, the large muscle should be cut between the lower valve and the body of the oyster. If we remove the lower valve, the knife will follow the curve of the shell, and will not be so likely to injure the soft parts. The teacher has an oyster with the muscle still uncut.

The oyster shell has two valves, one larger and thicker than the other. It is convex on one long edge, and concave or nearly straight on the other. It is broad at one end, and the beak is at the opposite narrow end. The deposits of lime are much more uneven than on the clam shell, and the shell is strongly roughened along the lines of growth. One shell has a piece of rock on the thick valve, another has a young oyster fastened to it, and a third has a piece split off from its large valve. These teach us that the oyster was attached to some object by the large, thick valve, which is therefore the lower valve. The upper valve is smaller and thinner than the lower one. To find the right and left valves, after holding a clam shell in the proper position, we hold the oyster shell in the same way, with the beak pointing away from us and the concave side uppermost. Then the right valve is on our right side, and the left valve on our left.

Various substances are found on the outside of the shell. Something red proves to be a bit of red sponge. Twisted white tubes were made by tube-building worms. Some little mats with pinholes in them are the skeletons of tiny creatures, one of which lived in each hole, or cell. Finally, some of the shells have been partially riddled by the boring sponge, which makes its tunnels by dissolving away the lime.

The teacher now cuts the muscle in her oyster, and as the children follow the course of the knife in her hand, they see that there is but one muscle, and that after cutting that, the shell may be opened by breaking the ligament.

The hinge is not exactly at the beak, but a little distance from it, and between the two is a widening groove containing some dried remains of ligament (Fig. 1, *l*). In the young oyster the valves must have been united at the beak, so the groove is the downward path of the ligament, as the oyster grew. The inside of the valve is white. There is a large, dark impression near the center of the valve. A line extends nearly around the shell, not far from the margin. This is the pallial line. By laying the valve back over the oyster we see that the large dark spot fits over the muscle, and hence is the muscle scar.

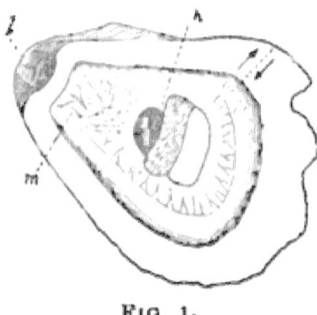

FIG. 1.

This scar has also traveled downward from the beak. To show this, file the outside of the thick valve to some depth, and strike it sharply with a hammer, when the inner layers of the shell will split off, disclosing the track of the muscle.

The oyster has a mantle that is open all around. The mantle is only a little thickened at the edges. The edges of the mantle are joined together in just one place, the bar (Fig. 1, *b*), where the convex and the concave sides of the shell meet. In front of the muscle is a clear space containing the heart (Fig. 1, *h*), a little whitish bag. Under the edge of the mantle are the gills, two pairs, as in the clam. Near the beak on the convex side of the shell are the two pairs of palpi, and between them the mouth (Fig. 1, *m*).

Lesson II.

This lesson, which is mainly a review, may be given with the aid of the shells, according to the following outline:

Tell me some things about the oyster shell.

It has two valves. One valve is large and convex, the other is smaller and flat. The large valve was fastened to a rock. The shell is broad at one end and pointed at the other. The pointed end is called the beak. The outside of the shell is very rough, and the lines of growth show very plainly. There is a hinge not far from the beak, and a brown ligament.

What can you tell about the inside of the shell?

The inside of the shell is nearly smooth. It is white or yellowish. Near the middle is a large dark place made by the muscle. There is a pallial line near the edge.

The shape of the gills and the palpi may also sometimes be seen on the inside of the shell.

In what ways are the oyster shell and the clam shell alike?

Each has two valves. They have each a hinge and a ligament. Each has a beak. They have a pallial line. Each has a brown skin and lines of growth on the outside. They are made of layers of lime and flesh.

In what ways are they unlike?

The clam shell is smoother than the oyster shell. The oyster shell has the beak at one end, the clam shell has it on top. The clam shell has both valves of the same size, but the oyster has one large valve and one small one.

If the children are old enough, they may now be led to see that while the left side of the mantle of the oyster can work steadily at shell-building, the right side is constantly interrupted in its work by the opening and closing of the valves, and that this accounts for the small size of the right valve.

Tell me in what ways the soft parts of the oyster are like those of the clam.

They have each a mantle and two pairs of gills. They have a mouth and two pairs of palpi. They have a heart.

Tell me some things that are unlike in the soft parts.

The oyster's mantle is open, and the clam's is closed. The oyster has one muscle, and the clam has two. The oyster's heart is close to the muscle, and the clam's is under the beak. I don't see why the oyster hasn't any foot nor any siphon?

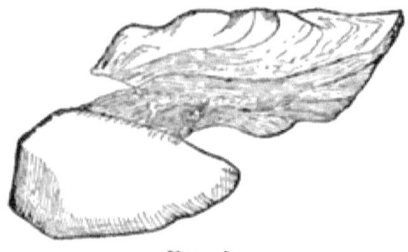

FIG. 2.

Where does the clam live?

He buries himself in the mud. He digs down into the mud with his foot.

What use has he for his siphon?

He reaches up to the water with it.

Where does the oyster live?

He fastens himself to a rock. I see; he doesn't need a foot because he doesn't dig in the mud, and he doesn't need a siphon because his whole shell is in the water.

Oysters naturally live on rocks or hard substances (Fig. 2), and after the young ones swim about for a while, they die if they cannot find something hard to grow on, but they fatten better for the market on muddy bottoms, where there are great quantities of tiny plants for their food. So when they are half grown, the oystermen take them up and "plant" them on the mud in some warm bay or at the mouth of a river, where they are left for a year or two. But they never dig in the mud, and so need neither foot nor siphon.

I should think the currents of water would get mixed if there isn't any siphon.

Is there any place where the edges of the mantle are joined together?

Yes, at the bar.

Then what is the use of the bar?

To separate the two currents of water.

One current flows in under the convex border of the mantle, passing over and through the gills, and carrying food to the mouth, the other flows out on the opposite side of the bar, as indicated by the arrows in Fig. 1.

As in the case of the clam and oyster, so with other mollusks the presence or absence of the foot and siphon is a sure guide to the habits of the animal.

THE SNAIL.

Lesson I.

The French edible snail is the one we use if we can get it, but a lesson can be given on any good-sized land or water snail. The slug (Fig. 1), which is simply a snail without any external shell,

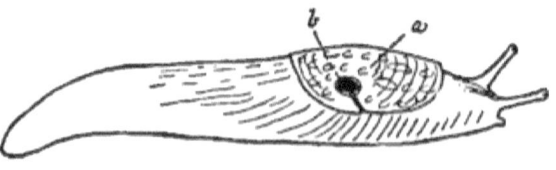

Fig. 1.

can be obtained at any season, it is said, by setting a trap for it in a greenhouse. The trap is a box of moist bran, which will attract it as cheese allures mice. But the slug is so likely to be repulsive to pupils that it is much better to collect snails in the summer and keep them in flower-pots or boxes of earth covered with moss to retain the moisture. If kept in a cool place, they will hide away under the moss, close up their shells with a layer of mucus, and sleep comfortably through the winter. If any of them come out occasionally on warm days, they will like to be fed with some wet bran or Indian meal. When they are active, the box should be covered with a wire netting to prevent their escape from such narrow quarters. A large colony has been kept in this way for more than a year, fed during the summer upon lettuce, of which they are very fond.

If the lesson is given in the winter and the snails have been kept in a cool place, the shells will be closed with a layer of mucus, and we can study them before the snails come out.

The shape of the shell reminds us of "a horn curled up." It has often made five or six turns in coiling, and each turn is called a whorl (Fig. 2). The whorls together form the spire, and the lines between the whorls are the sutures. From the opening, or aperture, we trace the whorls up to the apex, where we find the little bag-like shell that covered the baby snail. Then we follow the whorls from the apex to the aperture, and watch them grow larger at every turn. We remember the lines of growth on the clam and oyster shells, and decide that the delicate lines running parallel with the edge of the aperture must be the snail's lines of growth.

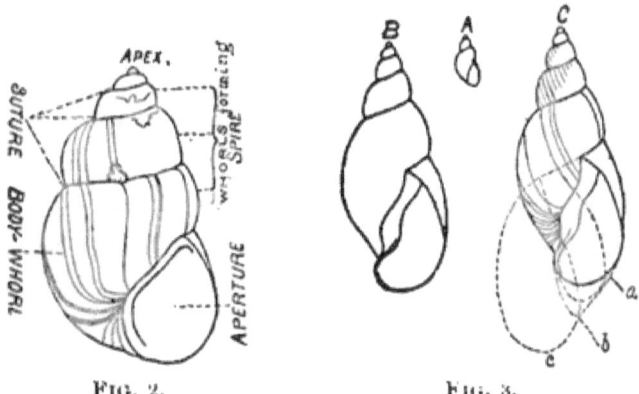

Fig. 3 shows how a shell grows. *A* is a young shell; *B*, the same, fully grown; and *C* is the same as *B*, but with the lines of growth represented on it. The dotted lines on *C* show the way in which it grows, *a* representing the first added layer, *b* the second, and *c* the appearance of the shell when another half-whorl has been formed. Some of our shells have a layer of thin and very brittle shell around the aperture, evidently just formed and not yet hardened.

Some snails' eggs, or very young snails, will add interest to this part of the lesson. In spring and summer the eggs of the pond snail may be found on the under surface of the leaves of waterplants in masses of jelly, in which the eggs look like little dots. These will soon hatch out, if kept in a jar of water. The shells of young snails are transparent, and the growth of the body and the gradual coiling of the shell can be easily watched.

Holding the shell in the hand with the opening toward us and the apex uppermost, the aperture is toward the right hand. It is usually on the right, so when we find one on the left, we will be careful to keep the shell.

Our snails have closed up the aperture of their shells

by making a door of the mucus, or slime, that comes from their foot. Every year when cold weather comes, the land snails hide away under the roots of a tree, or under a log or stone, and make this door to shut themselves in. Most of the water snails have a door all the time, which they can shut whenever they please.

Lesson II.

Just before the lesson we drop our snails into a bowl of warm water, and they are soon coming out, foot foremost.

Holding them upside down, we see a fleshy rim around the aperture. This is the mantle, which builds the shell.

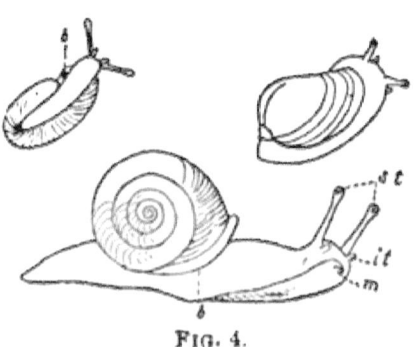

Fig. 4.

Just under the edge of the shell is the breathing-hole (Fig. 4, *b*), which keeps opening and shutting, and leads into the snail's simple lung, only a sac in the mantle with blood-vessels in its sides. The breathing-hole is on the right side when the aperture points toward the right, and on the left when that points to the left.

Two pairs of horns, that are gradually pushed out, are the tentacles (Fig. 4, *s t* and *i t*). The long tentacles are the eyestalks; the others are used only as feelers. We touch a long tentacle, and a black thread pulls the eye down inside. The black thread is the lining of the tentacle, which contains muscles that draw the end in just as we turn the finger of a glove when we pull it off in a hurry.

The snail moves on its broad foot, and by putting it on

a piece of glass we see that the foot is a large sucker that moves in little waves. The glass is soon covered with slime, poured out of a very small opening in the bottom of the foot near the head. In this way the snail coats bits of earth and stone with a smooth

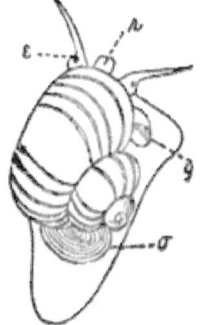

FIG. 5.   FIG. 6.

glaze, that prevents them from irritating its soft foot.

Finally, we feed the snails, and watch the hard brown jaw (Fig. 5) as it bites off pieces of the young leaves of cabbage, lettuce, or celery.

FIG. 7.

These pieces are chewed by the teeth on the snail's tongue, which point backward when the tongue is drawn in, but are made to stand erect when the snail is eating and the tongue is pulled forward by muscles. As the tongue works backward and forward, the teeth grind the food against the hard jaw, and also the cartilage that lines the upper part and sides of the mouth. As many of these tiny teeth are worn out every time that the snail eats, new ones are continually growing in a little pocket behind the tongue and pushing forward to take the place of the old ones.

If snails and slugs are troublesome in gardens, they may be killed by sprinkling dust or ashes wherever they are. They throw out so much mucus in their efforts to remove the dry particles that they soon exhaust their strength and die of weakness. In Eu-

FIG. 8.

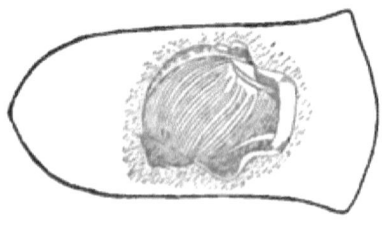

FIG. 9.

rope, where they do so much mischief in the vineyards, the people take their revenge by eating them in turn.

Pond snails have but one pair of tentacles, at the base of which are the eyes. Some of them, like the one seen in Fig. 6, have a horny scale, or operculum, which closes the aperture of their shells, and the mouth on the end of a rostrum, or beak, and also breathe by gills. Others (Fig. 7) have no operculum and no rostrum, and are air-breathers.

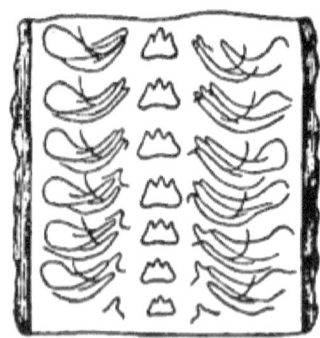

FIG. 10.

At the seashore we shall find the large *Lunatia* (Fig. 8). Fig. 9 shows this as it crawls about partly buried in the sand, with its broad foot extended and the shell almost covered by the soft body. With the teeth on its tongue (Fig. 10) it drills into the shells of other mollusks and eats out their bodies through the hole. This snail lays its eggs in the "sand collars" (Fig. 11) we have so often picked up on the beach, the round spots on them being egg-cases, and each one containing several eggs.

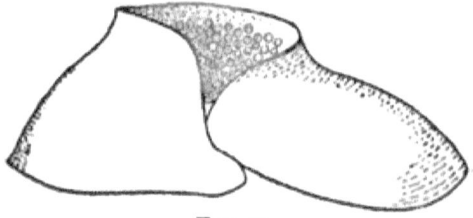

FIG. 11.

THE EARTHWORM.

To rouse herself to the proper pitch of enthusiasm on this subject, the teacher has only to read Darwin's *Vegetable Mould and Earthworms*. The most inveterate prejudice against everything that crawls will be unable to survive a careful study of this book.

Our living earthworms are kept for winter study in a box of earth in the cellar, covered to prevent their escape, but not so closely as to exclude the air. The earth must be kept damp, as worms breathe only through the skin, which must therefore be always moist. Occasionally they will need a few leaves for food, which will disappear by degrees inside their burrows. At night, when the worms are busy foraging, if we remove the cover quietly, and are careful not to let the light fall directly on them, we can surprise quite an active colony, each one exploring the earth around him with the forward end of the body, while the tail still clings just inside the burrow.

For our lesson each child has a living earthworm in a little dish. Half of the dishes are filled with water, half with earth. A little talk about the habits of the worms occupies a few minutes, and gives the children time to become accustomed to their movements and ready to observe them.

At first both ends of the long, ringed body look alike to us, but by degrees we discover that the end which always moves forward (Fig. 1, *l* 1) is pointed and terminates in a little knob, the upper lip. There

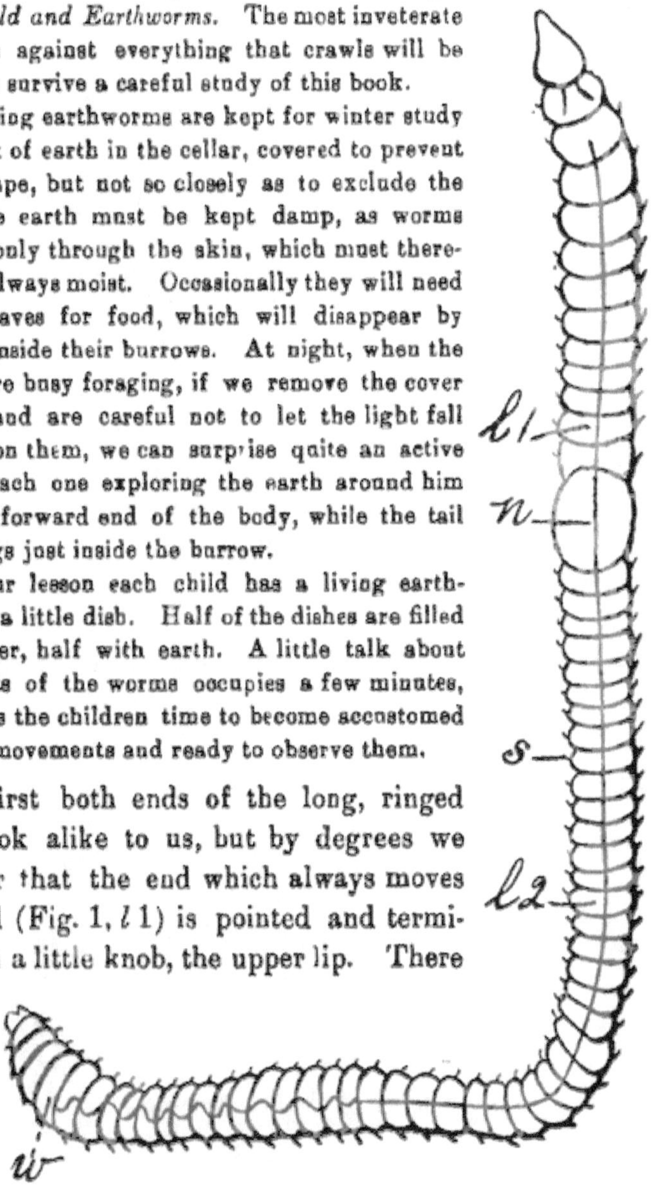

FIG. 1.

is no distinct head, but the mouth is at this end. Here, too, are the largest rings. The other end, or tail (Fig. 1. *l* 2), consists of smaller rings, and is flattened at the tip (Fig. 1, *w*). About one third of the distance from the head to the tail is the saddle (Fig. 1, *n*), where the rings are thickened and show much less plainly.

We can count the rings on an alcoholic specimen, but we must remember that the surface of each ring is divided by a skin fold, so that there are only half as many rings as folds. Extending inward to the intestine are muscular partitions which separate the rings and divide the body into a series of chambers.

In color the worm is reddish brown, the upper side being much darker than the lower. At first we think it has no skeleton, but if we put an alcoholic specimen in water for a few hours, we can remove a thin, horny cuticle, or outer layer of the skin. Sometimes we detect a beautiful iridescence, caused by the play of light on the tiny folds of the cuticle.

Now, taking the head end of the worm gently between the fingers, and holding it up, we shall see it open its mouth in its efforts to escape, and thrust out a large, white, membranous pouch, into which the mouth leads. This pouch is drawn forward when the worm is digging its burrow, thus swelling out the head and making a much larger opening in the soil. As we hold the worm, we feel the pressure of the stiff hairs, or bristles, on its sides, which we see as bright points when it draws up its rings and pushes against our fingers. There are two double rows of these hook-like bristles (Fig. 1, *s*) on each side of the body, just where the darker color of the back joins the lighter color of the under side.

Earthworms have no teeth, but with their lips they pinch off the soft parts of decaying leaves, on which they feed. The red thread along the worm's back is the principal blood-vessel. The dark tube passing through the body is the intestine, which contains earth, swallowed partly to get it out of the burrow, and partly for the decaying plants in it.

THE LOBSTER.

Lesson I.

A single lobster is killed for our lesson by keeping it in warm water for a few hours. Many pupils will hardly recognize it in its natural dark-green coat, while others will not only know it, but will also describe the curious lobster-pots in which it is taken on our sea coast. In the interior of the country, where crayfishes abound, the lobster is not needed, though it is an excellent plan for the teacher to have one while the pupils have crayfishes. The proper position for our specimen is with the back uppermost and the head pointing away from us. As the lobster is held up in this position before the class, some observations will be quickly made:

Its color is dark green, and reddish on the claws. Its body is shaped like a tube. It is covered with a hard crust. The crust is its skeleton. The body has two parts, the head and the tail. The tail is made of rings. The head is covered with a great shell like a saddle.

The part that has been called the lobster's head includes his chest as well, and we put on the blackboard the proper names of the parts mentioned, in this way:

The two parts of the lobster's body are the head-thorax (Fig. 1, *cth*) and the abdomen (Fig. 1, *ab*). The large shield that covers the head-thorax is the carapace.

But these two parts are not the whole lobster. What else has he?

He has legs, claws, feelers, and eyes. He has little flaps on the rings of the abdomen. He has a sharp nose between his eyes.

We find that the "sharp nose" is only the pointed end of the carapace, and is called the beak. We wonder what it is there for. To protect the eyes, of course. But the legs and feelers, which are attached to the body by joints, are all called appendages. These appendages are not all in one piece like the bristles of the earthworm, but are themselves jointed. For the first time we begin to study animals with *jointed appendages*.

We see that the abdomen consists of six rings and a flattened piece at the end called the telson (Fig. 1, *t*).

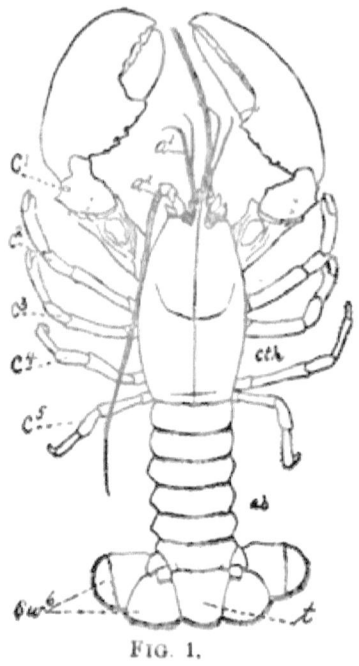

FIG. 1.

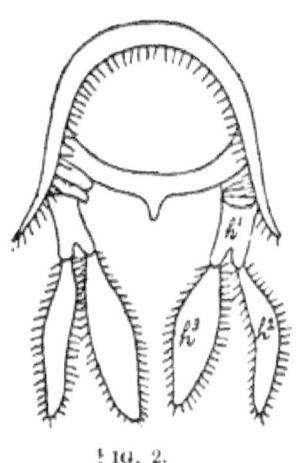

FIG. 2.

It is not to be wondered at if the tail fin puzzles the class, with the two broad lobes on each side of the telson, while in the crayfish the telson itself is jointed. But after careful observation they will see that the telson is a part of the body, while the lobes on either side (Fig. 1, sw^6) are parts of organs that are jointed to the body, that is, parts of appendages. If this is clearly seen, the next question will be correctly answered.

How many appendages can we find on any ring of the abdomen?

There are two on each one. There is a pair of appendages on each of the six rings, but none on the telson.

Are these appendages all alike?

The front pair are very small and the hind pair very large, but the others are nearly the same size.

These appendages are all used in swimming when the lobster is young, so are called swimmerets or little swimmers. In the breeding season the female carries her eggs glued to the small swimmerets.

Those who have seen a live lobster try to get away in a hurry, know why the sixth pair of swimmerets are so large. Their broad lobes spread out on each side of the telson in such a way as to make a powerful tail fin with which the lobster strikes the water, thus sending himself forcibly backward.

We examine the third pair of swimmerets, and find that they consist of a stem (Fig. 2, h^1) bearing two flattened lobes (Fig. 2, h^2 and h^3).

Some of the class will doubtless observe that the second, third, fourth, and fifth pairs of swimmerets are all made on the same plan, but it is not wise to force any such comparison upon grammar school pupils. When they are older it will be a part of their work to trace one common plan in all its variations through the whole series of appendages, but such study of homologies is for maturer pupils than ours.

Fig. 2.—Third segment of the abdomen with its pair of appendages.

Lesson II.

In order to give the class a correct idea of all the appendages, we make blackboard sketches of the mouth-parts (Fig. 3), and also sew the mouth-parts, the antennæ, and the eyestalks to a piece of dark-colored pasteboard.

Review of Lesson I.—The lobster is dark green, with reddish claws. It is tubular in shape. Its skeleton is a hard crust. The body is in two parts, — the head-thorax and the abdomen. The head-thorax is covered by a great shield called the carapace. The lobster has many pairs of jointed appendages. The abdomen has six rings and a flat telson. There are six pairs of swimmerets on the abdomen. With the small swimmerets the lobster carries its eggs; with the pair of large swimmerets and the telson it takes long jumps backward.

Biting Jaw,	Mandible, ...	
Little Jaws, or Accessory Jaws, or Maxillæ,	{ First Maxilla, . . . { Second Maxilla, . .	
Foot Jaws, or Maxillipeds,	{ First Maxilliped, . . { Second Maxilliped, . . { Third Maxilliped, . .	

FIG. 3.

Outline of new work. — We will count the appendages of the head-thorax. There are five pairs of legs (Fig. 1, $c^1 - c^5$), and one pair end in great claws. There are two pairs of little legs. There are "lots of little legs and things" next to the big claws.

All these little legs can be seen fastened to this piece of pasteboard How many pairs are there?

There are two pairs of little legs. There are three pairs that are split, and look something like swimmerets. There are two great teeth, with a pair of tiny legs fastened to them.

The three pairs next the great claws are the jaw-feet, called also foot-jaws or maxillipeds. The next two pairs are little jaws or maxillæ, and the "two great teeth" are the pair of mandibles or chewing jaws. Let us put our probes between the mandibles into the mouth. These six pairs of appendages are called mouth-parts. What other appendages has the lobster?

He has two pairs of feelers. Those are his antennæ.

What is curious about his eyes?

They stick out from his head, and can be moved about.

The eyestalks are another pair of appendages. How many pairs of appendages are there on the head-thorax?

There are fourteen pairs.

How many pairs on the abdomen?

There are six pairs.

Now we will compare the legs. How many joints have they?

The big legs have six joints and the others have seven. The joints move in different ways. The first pair of legs end in large claws, the next two pairs in small claws, and the last two pairs in sharp points. One great claw has broad, blunt teeth, the other has sharp teeth.

Since the lobster lives in shallow water, he uses his four pairs of small legs in walking over the rocky bottom. The great claws are kept for fighting and tearing his prey. If he loses one in a duel, another soon takes its place. While the broad teeth on one claw often anchor the lobster to some large seaweed, or are used as millstones for crushing his food, the other claw catches fish and tears them apart with its sharp teeth.

While the great claws capture moving prey, the third pair of jaw-feet pick up food from the bottom, and their saw-like inner edges help to tear it in pieces. The other mouth-parts, especially the strong mandibles, do the rest of the work of biting and chewing, though the little jaws seem too soft to be of much use.

We notice that the first pair of antennæ (Fig. 1, a^1,) are very short and have two parts, but the second pair (a^2) are very long and made of many little joints.

The ears are in the lower joint of the small antennæ, which is flattened on the upper side and surrounded by hairs. If these are pushed apart, a small, clear, oval space is seen, which is the outer covering of the ear.

Pupils who think a lobster needs eyes in the back of his head, since he takes his flying leaps backwards, see how this need is met by the movable eyestalks, which enable him to turn his eyes in any direction.

Lesson III.

Review of Lessons I. and II.

The following list of the lobster's appendages is put on the blackboard, and the children are asked to tell all that they can about each pair:—

Head-thorax.	1 pair of eyestalks. 2 pairs of antennæ. 1 pair of mandibles. 2 pairs of little jaws. 3 pairs of jaw-feet. 5 pairs of walking-legs.
Abdomen.	6 pairs of swimmerets.

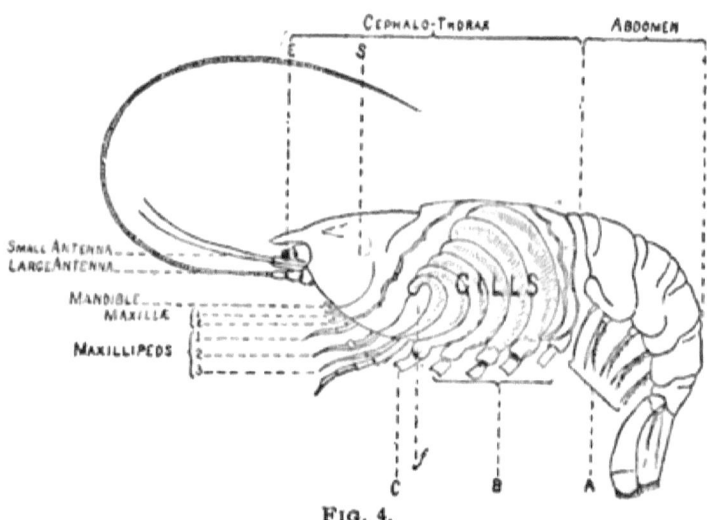

Fig. 4.

For young children these appendages may be simply classified as eyestalks, feelers, chewing feet, walking feet, and swimming feet.

OUTLINE OF NEW WORK.

The abdomen consists of rings, and we find but one pair of swimmerets on a ring. Now let us see whether the appendages of the head-thorax are borne on rings, too.

With strong scissors we cut along the groove of the carapace, from the front half-way up the back, and bend back the carapace, as shown in Fig. 4, thus displaying the plumy gills, some fastened to the sides of the thorax, others to the legs and the jaw-feet. Cutting away the gills, we see a thin white shell, which seems to be made of several pieces that have grown together. Now, by lifting each leg in turn, we see that every pair of legs is borne on one of these pieces, which are therefore the lower portions of several rings. As we approach the mouth and the appendages become more crowded, the rings are completely grown together, but we conclude from what we have seen that each pair of appendages represents a ring, and the head-thorax therefore consists of fourteen rings soldered together.

Children must not be hurried to this conclusion, but led up to it slowly and carefully by patient questioning. They will *know* only what they discover for themselves.

We have found the gills safely hidden away under the two sides of the carapace, in the lobster's vest-pockets, which are open at each end, so that the water may rush out as well as find its way in. But as there must be always a current of water over the gills, a spoon-shaped organ,—the gill-scoop or flabellum (f, Fig. 4),—attached to each of the second maxillæ, is constantly scooping it out at the front of the pockets as fast as it rushes in at the back. In this way a fresh supply of oxygen is continually brought to the blood in the gills.

A probe put into the mouth passes into the stomach, a

wonderful machine in the top of the head. As if the lobster had not jaws enough for tearing and chewing, he has teeth inside his stomach, by which his food is so finely ground that it will pass through a strainer of little hairs into the back part of the stomach.

The heart, a six-sided, spongy organ, which sends the purified blood from the gills over the whole body, lies under the carapace just behind the transverse groove on the back. We have now seen that the carapace covers and protects the most important part of the lobster's body, that containing the gills, the stomach, and the heart.

Very young lobsters are beautiful, transparent little creatures, that swim about at the surface of the water, and are quite different from their parents. Fig. 5 represents one magnified, while *A* shows its natural size.

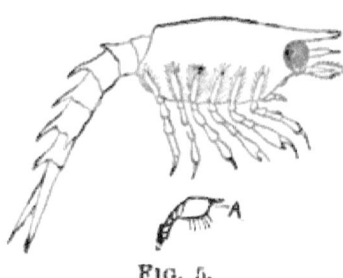

FIG. 5.

A lobster's shell once formed never increases in size, and like a boy's jacket, must be thrown away when he has outgrown it. This is the way the lobster pulls himself out of it: First, the shell splits here on the back between the head-thorax and the abdomen, or down the middle of the head-thorax, or sometimes in both places; then the poor fellow wriggles and twists till he gets his head and legs out of their sheath, when with one pull he frees the rest of his body. The new shell is formed before the old one comes off, but is very soft and loose, so that the body can grow to fill it. This shell hardens and thickens in a few days, but in this short time the lobster has grown as much as he will until he moults again.

THE CRAYFISH AND THE CRAB.

Though the greater part of the lessons on the lobster applies equally well to the crayfish (Fig. 1), still, for the sake of those who will make use of the latter alone, a few more points may well be brought out here.

The crab (Fig. 2) deserves our attention not only because it is so common and easily obtained, but also because it furnishes a striking instance that children can appreciate of the changes in an animal brought about by change of habit, even when the general plan of structure remains the same. Those who can procure both the lobster and the crayfish may devote one lesson to a comparison of the two as a review of this important type of structure, when children's quick eyes will discover many minute points of resemblance

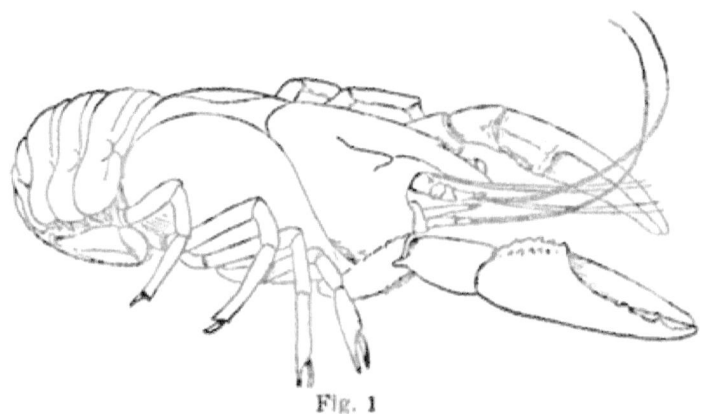

Fig. 1

or difference that it is unnecessary to touch upon here. It will of course be impossible in practice to combine the crayfish and the crab in one lesson, as is done in this outline.

The home of the crayfish is in fresh-water streams, where it is most active towards evening, seeking shelter from the heat and sunshine of the day under the shade of stones and banks. In the winter it burrows in the banks, not to sleep, however, for on warm days it lies at the mouth of its burrow watching for food. Like its salt-water cousin, it is not at all fastidious, but greedily swallows frogs, tadpoles, water-snails,—anything and everything that comes along.

A little colony of crayfishes can be kept in a cellar or any cool, dark place, in a tub with two or three inches of water and earth, and stones enough for a false river-bed. They will show considerable ingenuity in building tiny caves, into which they beat a headlong retreat when disturbed, and which they will rebuild as often as we demolish them. Unfortunately, they are cannibals, and even if well fed with raw meat, show a wicked preference for their own brothers and sisters, that numbers the days of our little settlement.

The dull green or brown color of the crayfish, its smaller size than the lobster, its similarity in shape, skeleton, and appendages, and the equal number of appendages, are all noted.

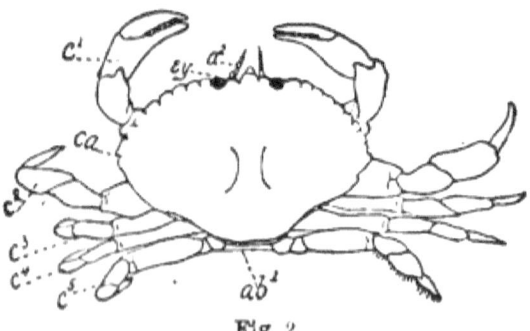

Fig. 2.

A preparation of the mouth-parts upon pasteboard is almost essential in the case of the crayfish, and it will be found well worth the while to glue the carapace and the rings of the abdomen to the center of the card, and then to arrange all the appendages in regular order on either side.

The division of the telson into two jointed pieces, the variation in the anterior pairs of swimmerets, the different shape of the rostrum, the lower part of the long antennæ, and the great claws, should then be observed.

In studying the crab, the *differences* between that and the lobster at first engross our attention.

It is nearly all head-thorax. The abdomen is very small, and is tucked up under the thorax. The carapace is broader than it is long. The eyes are nearly hidden in two hollows under the edge of the carapace. The carapace has no beak, and the antennæ are very small. The great claws are bent like two arms, so that the crab can bring them up to his mouth. All the small walking-legs end in points. The mouth-parts are covered by two plates.

These two plates are really the enlarged joints of the third pair of jaw-feet.

Turning back the abdomen, we find nothing that can be called a swimmeret, but either two or four pairs of appendages, used by the female in carrying the eggs. The abdomen, no longer used in swimming, has become as small as possible, and is safely kept out of the way.

But the crab is like the lobster in some respects. It has head-thorax and abdomen. Its skeleton is a hard shell. It has five pairs of legs, and one pair has great claws. It has some mouth-parts. It has two pairs of antennæ. It has eyes on stalks.

By removing the under side of the carapace, which covers the gills, and the third pair of jaw-feet, which hide the other mouth-parts, we complete the proof that the number of appendages on the head-thorax and the whole plan of structure are identical with what we are so familiar with in the lobster.

THE HERMIT-CRAB.

A jar of hermit-crabs in alcohol will furnish material for a delightful lesson. If but few of the children have seen them in their homes, we begin with a short account of their habits.

Last summer I spent part of the vacation at the seashore, and one day, while I was walking on the beach at low tide, I caught sight of some snail shells moving along just below the water's edge. But who ever saw snails crawl so fast? And they were *on* the sand instead of partly

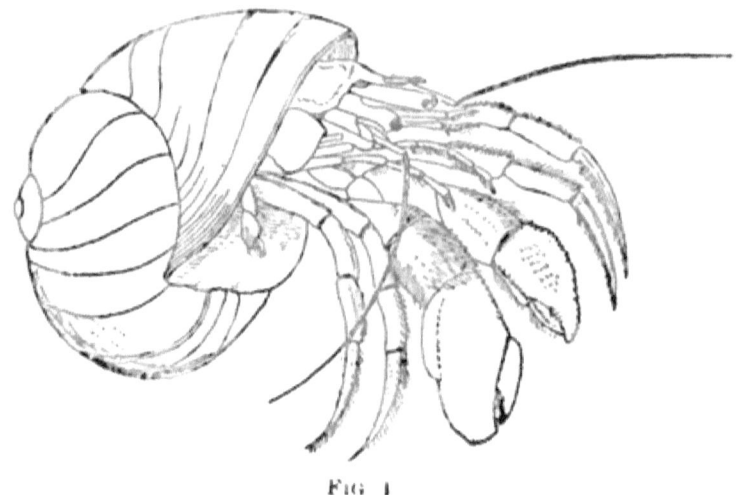

FIG. 1

buried in it. As I stooped to pick up one, I was just in time to catch a glimpse of a small lobster-like head and claws disappearing inside, while one of the big claws closed the aperture of the shell with a sharp click. I had come upon a colony of hermit-crabs (Fig. 1). I put some in a pail of sea-water and carried them to the house, where I could watch them. As soon as the water in the pail began to grow impure, the crabs left their shells entirely, and then I saw why they needed their houses. But when I gave them a fresh supply of sea-water, how delighted they were to crawl back to their shells, trying first one and then another till they found one of the right size, and then backing contentedly into it! Now let us examine our specimens.

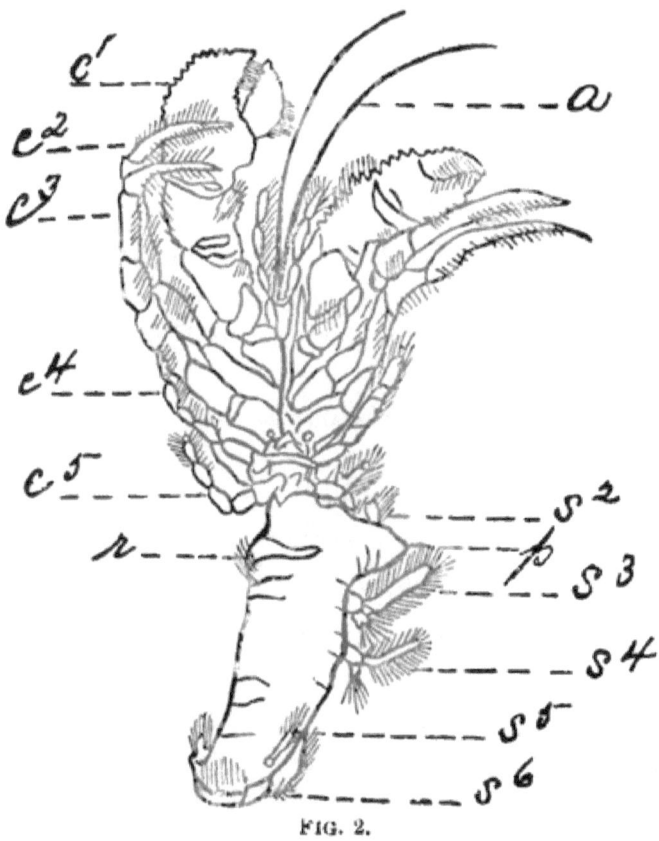

Fig. 2.

The body is nearly all soft. It has a shell on the head and claws and a little shield over the heart, but the abdomen is so soft that it almost breaks in two.

Why does the hermit-crab carry a snail-shell on his back?

To protect his soft body.

Has he the same appendages as the lobster?

He has two eyestalks, longer than the lobster's. He has two pairs of antennæ. (Fig. 2, a, the long antennæ.) He has one pair of hard mandibles. I think he has just as many mouth-parts as the lobster.

You are right. He has two pairs of little jaws and three pairs of jaw-feet.

He has one pair of great claws (Fig. 2, c^1) and four other pairs of walking-legs (Fig. 2, c^2–c^5). One big claw is larger than the other.

<small>Most of the class will find that the right claw is larger than the left, because the aperture of most snail-shells is on the right side, and this claw is used to close it up when the crab is inside.</small>

The walking-legs are different from the lobster's. The second and third pairs are long and pointed at the end, and the fourth and fifth pairs are short and end in little claws.

The last two pairs are probably used only for holding the creature in the shell.

Are there any swimmerets?

There are three or four on the left side (Fig. 2, s^2–s^5), and one pair at the end of the abdomen (s^6).

<small>The male has but three appendages on the left side of the abdomen, but the female has four, which are longer than those of the male because used in carrying the eggs. Both have the pair at the end, used as claspers. The hard crust on this pair and on the last two segments of the abdomen, a hard ridge (Fig. 2, r) on the under side of the abdomen, and a projection (p) close to it on the left side, all help to hold the crab in his shell.

In looking for the gills, we trace the flap that covers them up to the side of the body, where it shows beautifully that it is nothing but a double fold of skin, thus making clear to teacher, if not to pupils, its real structure in the lobster.

One can hardly find stronger proof than in the hermit-crab that an animal becomes adapted to its surroundings, and also that parts unused will either disappear or so far degenerate as to be incapable of use.</small>

THE BEACH-FLEA.

We may choose either the Gammarus (Fig. 1), colored like the heavy masses of dark green seaweed under which it hides, or the equally nimble Orchestia, so precisely like the sea-sand in its color that many a one escapes from us as we dig into their holes on the beach. The Gammarus is the one first described in this lesson.

Observations to be made: The beach-flea has a very narrow body, strongly curved for jumping. The rings can be seen on the whole of its body except the head.

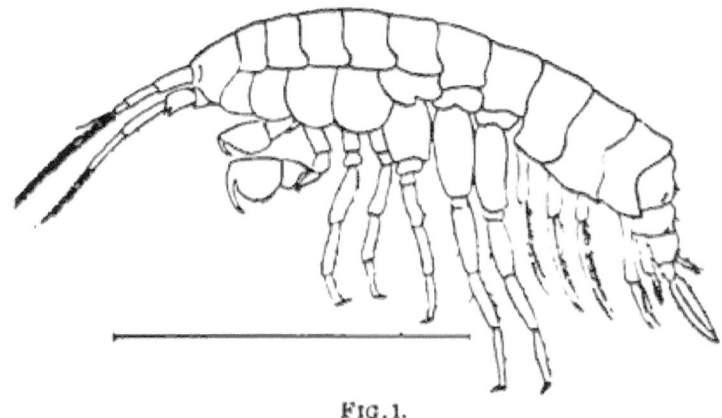

FIG. 1.

The last four rings of the abdomen are narrower and harder than the others. There are seven pairs of legs in all. Two pairs of these have great claws. There are six pairs of swimmerets. The first three pairs of swimmerets are soft, the others are hard because used in jumping. There are two pairs of antennæ. The eyes are like two curved black lines, and are not on stalks.

An interesting comparison may be made between the shape of the body of Gammarus and that of Orchestia. The former has so narrow a body that it cannot stand upright, suited to the crevices through which it must often make its way; while the latter, which burrows in the yielding sand, has a much broader thorax, and can keep its balance in an upright position. Gammarus, however, must always swim on its side or back.

The beach-flea has no carapace covering both head and thorax. Remembering that the thorax is that part of the body which bears the legs, we count seven rings of the thorax, each bearing a pair of legs. The first ring of the thorax, that nearest the head, is so small that we must take it on trust, but bears a pair of jaw-feet, nevertheless. The first pair of legs end in rude pincers formed by bending the little hooked claw backward toward the next joint above, which is somewhat broadened. The second pair have larger pincers of the same sort, with the joint above the last much more broadened. The remaining five pairs all end in single claws. The pear-shaped bags at the base of the legs are the gills.

We are careful to say only that we *count* seven rings of the thorax, for we know it really consists of eight, even if we can see but seven.

The mouth-parts are so tiny that they can scarcely be seen without a magnifying glass. They consist of the pair of jaw-feet already mentioned, two pairs of maxillæ, and one pair of mandibles. The head also bears two pairs of antennæ nearly equal in length.

By placing the head under a magnifier and pressing back the jaw-feet with a pin or a dissecting needle, a careful worker may satisfy herself that the hard mandibles and the maxillæ, soft and leaf-like, are really present.

Now taking Orchestia and comparing it with Gammarus, we make the following discoveries:

The eyes are not on stalks, and are nearly round. The first pair of antennæ are very short, indeed. Plates of shell cover the legs just as in Gammarus. The last four rings of the abdomen are very small, and the last three pairs of swimmerets are hardened and much crowded. There is but one pair of pincers, and these are on the second pair of legs. The mouth-parts are the same as in Gammarus, but are more plainly seen.

There are no pincers on the first pair of legs, but the section above the last is turned downward and the terminal claw bent toward it, as if in the attempt to form pincers. An interesting series may be made, beginning with this rude suggestion of a grasping organ, placing next the first pair of pincers in Gammarus, then the second pair in Gammarus, and finally the second pair in Orchestia, in which the pincers have attained their greatest size.

THE SPIDER.

Lesson I.

The common large, brown, long-legged spider (Fig. 1), which spreads its web on plants and rests securely in the tube leading downward from it, is the one described in these lessons. We use alcoholic specimens. The large black and yellow field and garden spider is another excellent one for class use, and is so common that it is easy to collect a sufficient number. Still another good one is the great round-web spider, often found in barns, but its body is softer and more easily injured.

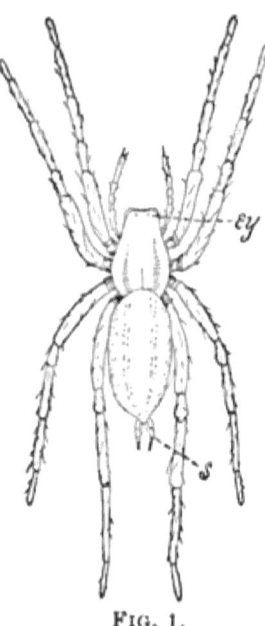

Fig. 1.

A small round-web spider has also been imprisoned in a box with a glass cover, where its movements could be watched. We have seen that it did not rest until it had spun in every direction through the box, so that it could go anywhere in its new home without stepping off the threads. When this was finished, it was fed with flies, and the process of killing them carefully observed, from the time when they were first bound with tiny ropes till they were let fall, dry and juiceless, to the bottom of the box.

The bodies of spiders are so soft and so easily broken that it is absolutely necessary to fasten them to bits of cork by a pin through the thorax if they are not to be ruined in the handling. A large pin and a bristle are also given to each child, since the probes are too clumsy for use upon such little creatures.

Facts Already Learned by Observation:

The spider in the box has spun threads across it in

The Spider.

every direction. It hangs upside down in its web. The spider tied up a fly in the web and held it there a long time, then it took off the threads and let the fly fall on the bottom of the box. The spider has four pairs of long legs. It has a big head and a round body with spots on it.

This last observation will soon be *corrected* from the alcoholic specimens. These must now be held with the back uppermost and the head pointing away from the pupil.

Outline of new discoveries:

The spider could be cut in halves by cutting lengthwise from front to back. The spider's body has two parts. Since the front part bears the legs, it must be the

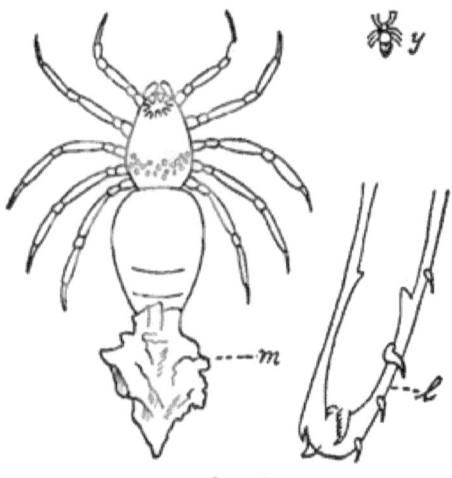

FIG 2.

head-thorax, and the other part is the abdomen. The abdomen is very soft and looks as if it might break off from the head-thorax. The skeleton is a horny crust covering the outside of the body. There is a dark band down each side of the head-thorax, and some dark stripes and spots on the abdomen. There is a little hollow near the hind part of the head-thorax.

This hollow shows where the large muscle that moves the sucking-stomach is attached. By the contraction of this and opposing muscles below, the top and bottom of the stomach are drawn apart, and the spider's liquid food is pumped backward from the mouth to the intestine.

The spider has one pair of short legs besides the four pairs of long ones. It uses the long legs for walking, and holds the others out in front like feelers. The first pair are not legs, but jointed feelers, called palpi.

In the young spider (Fig. 2) the young legs are shorter in proportion to the size of the body, and the palpi are evidently legs and are so used. Afterwards they do not lengthen as much as the others, gradually take a different position, and being used as feelers, are called *palpi*. The distinction between these and the legs needs to be clearly brought out, because we must think of spiders as having *only four pairs of legs*. Short palpi with broad and apparently distorted tips show that the specimen is a male.

Lesson II.

Review of Lesson I.—Spiders can spin webs. They catch insects in the webs for food. The body is in two parts, head-thorax and abdomen. The abdomen is large and connected with the head-thorax by a small joint. The spider has four pairs of long legs and one pair of palpi. The palpi are used as feelers.

The spider must now be held with its head toward the pupil.

There are eight eyes (see Fig. 1). Under the magnifier they look like tiny black beads. In front of the head are two clumsy things with little hooks on the end of them (Fig. 1). They look like short legs, and I can push them from side to side by pressing my

FIG. 3

pin in between them. I think the spider bites with them. They are its mandibles or biting-jaws. I can put the bristle into the mouth just between the mandibles. The spider does not need a large mouth because it only sucks the blood of the insects that it catches. On the under side of the thorax, close behind the mandibles, are the maxillæ or little jaws (Fig. 4, a). They are not separate appendages, but the flattened first joints of the palpi, which are used in chewing.

Figs. 3 and 4 represent the greatly magnified mandibles and maxillæ of a common garden spider, but not the one we are studying.

The appendages of the head-thorax are four pairs of legs, one pair of palpi, and one pair of mandibles.

When a spider bites, the poison is poured out through a tiny hole at the tip of each mandible. The poison sacs are partly in the head and partly in the upper joints of the mandibles.

The abdomen has no appendages but the spinnerets. In our spiders two of these are long and

FIG. 4.

stand out behind the abdomen like two tails (Fig. 1). Most spiders have three pairs of spinnerets shaped like so many knobs. By rubbing one of the hind feet over a spinneret a sticky fluid like white of egg is drawn out of all the little tubes in filaments, which instantly harden in the air. Hundreds of these strands unite to make the spider's thread, so delicate and yet so wonderfully strong.

Fig. 5 is one of the long spinnerets of our spider with the tiny spinning-tubes, *sp*, on the under side of the last joint. Fig. 6 shows the end of a spider's leg with the two-toothed claws, *o*, and the middle claw, *m*, without teeth, which is used as a thumb in holding and guiding the thread. These claws, as well as the toothed hairs, *t*, and all the other hairs on the leg, are so highly polished that it is impossible for a spider to be caught in her own web.

Just in front of the spinnerets is the small opening of the air-tubes, and further forward are two openings leading to the pair of air-sacs, in which the blood is purified as it passes through delicate membranous leaves.

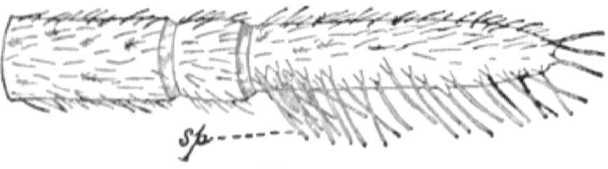

FIG. 5.

The eggs of spiders are laid in dainty cobweb cases often found under sticks and stones, or hung up in barns. Some spiders carry their young about on their backs until the little ones can look out for themselves.

Fig. 2 is a magnified view of a young spider just from the egg, with the first moult, *m*, still adhering to the end of the abdomen; *y* is the same spider, natural size; and *l*, the end of a leg greatly magnified to show an outer skin not yet shed.

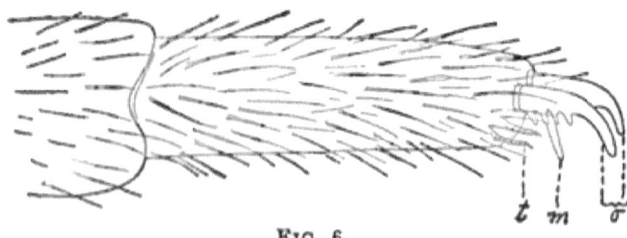

FIG. 6.

Many spiders live through the winter, hiding under fallen leaves and coming out the first warm day of spring.

Suggestions for Further Observation.—How does the round-web spider begin the web? In spinning the spiral, does she go from the center to the outside or from the outside to the center? Does the garden spider generally stay on her web? Why does the little gray jumping spider look so much more wide awake than others of the same size? Are there any spiders that are protected by their color?

THE GRASSHOPPER.

Lesson I.

The very largest grasshoppers we could find have been preserved in alcohol, and from a friend in Florida we have a small jar of the great "lubber grasshoppers," three or four inches long. Of our common ones, the flying grasshoppers make the best specimens, because they show so plainly the structure of the hind wings. We pin them to bits of cork, and also distribute pins before beginning the lesson. This insect is so familiar that the children are first allowed to tell what they know of its habits, and as many points of structure as they can discover for themselves, without much questioning. Some of the following observations are made:

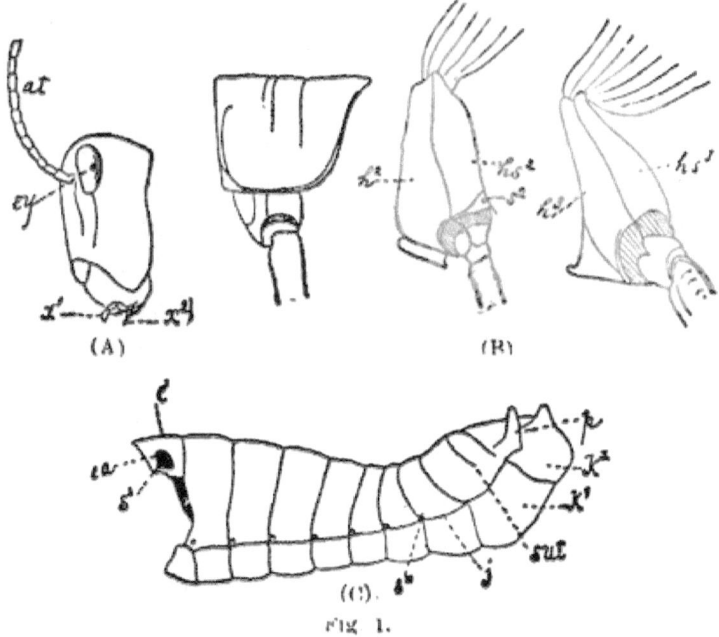

Fig 1.

The grasshopper has one pair of long legs and two pairs of short legs. It lives in the grass, and jumps with its long legs. It has two pairs of wings. It has a saddle on its back like the lobster. It has one pair of antennæ. It has two large eyes. It has a head-thorax and an abdomen.

Is there any movable joint between the lobster's head and its thorax?

No; the head-thorax is all in one piece.

If there is a joint between the grasshopper's head and its thorax, which bears the legs, then we will say it has a head and a thorax. Has the grasshopper a head-thorax or a head *and* a thorax?

It has a head and a thorax.

What are the three parts of the grasshopper's body?

The head, the thorax, and the abdomen (Fig. 1, A, B, C).

We observe next that the grasshopper has jointed appendages, that its skeleton is a horny crust on the outside of the body, and that it can be divided into two equal parts by a cut lengthwise from the head to the end of the abdomen. We stop a moment to recall other animals that we have studied whose bodies can be divided in halves in the same way.

An extremely careful count shows us that the abdomen consists of ten rings. The only appendages of the abdomen are three pairs of hooks, or "egg-points," at the end, forming the egg-layer. A large oval spot on the first ring (Fig. 1, *ea,*) is the ear, one on each side. We think this a curious place for the ears, but remember that the lobster carries its ears in its small antennæ.

The first ring of the abdomen (Fig. 1, c^1,) can be seen only on the back, as it does not reach wholly around the body, while in the male the ninth and tenth (Fig. 1, k^9, k^x,) are much larger on the under side. It is often said that the ovipositor consists of but two pairs of organs (Fig. 2, os^1 and os^2,), probably because the third pair (Fig. 2, os^3,) are very small and not readily seen unless the parts are distended. Neither are naturalists agreed as to the ears. They are certainly sense-organs, but may not be organs of hearing.

The collar behind the head at first almost deceives us into thinking it the whole of the thorax, but we remember just in time that the thorax bears all the legs, so this can be only the back of the first ring. On good speci-

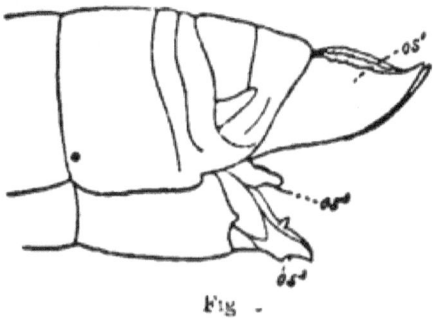

Fig -

mens, children can see that there are two rings behind this, both bearing legs. The thorax, then, has in all three rings, and bears three pairs of legs.

On the side, the second and third rings of the thorax, appear to be four rings instead of two. This is because the side of each ring consists of two plates (Fig. 1, B, h^2 and hs^2, h^3 and hs^3), which, in the typical ring lie one above the other, the upper one of which has here been forced out of its proper place until it lies behind the other.

The number of joints in the legs, the sharp spines with which they are armed, and the soft cushions padding the feet, will all interest the children, who will like to spend as much time on these points as can be well spared for them.

The second and third rings of the thorax bear the wings. The first pair of wings are straight and long. They meet on the back and cover the hind wings. The hind wings are broad and thin, and folded like a fan. In the flying grasshopper they are often beautifully colored. The hind wings are the ones chiefly used in flying, and the fore wings are made hard to protect them.

Lesson II.

The best way to learn the external anatomy of this or any other insect is to separate the body into its various parts, and arrange them on a card with the appendages in their proper places beside them, as shown in Fig. 1. Even if children cannot do this with perfect success, they will learn much in the attempt.

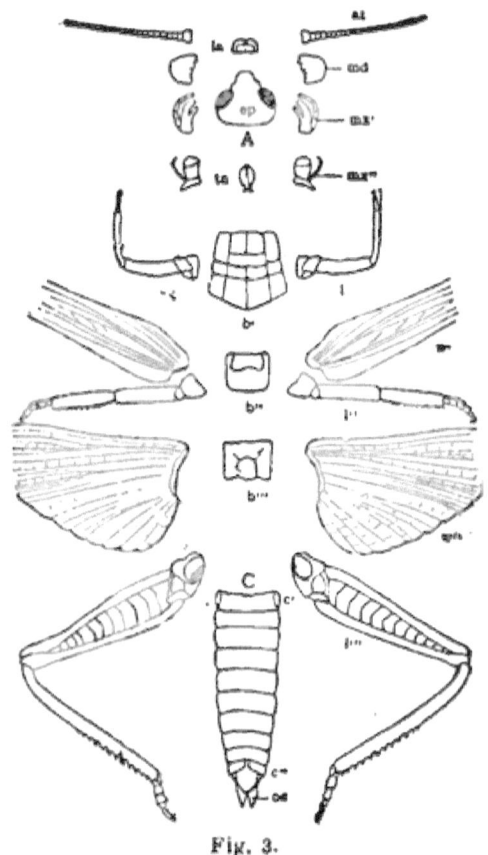

Fig. 3.

Review of Lesson I. — The grasshopper has head, thorax, and abdomen. Its skeleton is a horny crust that covers its body. It has jointed appendages. The thorax has three rings. On the thorax are three pairs of legs and two pairs of wings. The hind wings are folded like a fan, and the others are long and straight. The hind legs are long for jumping. The abdomen has ten rings and bears the egg-points. The ears are on the first ring of the abdomen.

The Grasshopper.

Outline of New Work. — The head is long and moves freely on the neck. There are two large eyes, one on each side of the head. The grasshopper has one pair of antennæ. In the center of the forehead is a simple eye (Fig. 2, *oc*,) that is easily seen, while two more simple eyes (Fig. 2, *oc¹*,) are placed, one on each side, in front of the compound eyes.

Holding the head of the grasshopper firmly between the fingers, and raising the loose flap that covers the mouth-parts, called the labrum, or upper lip, (Figs. 1 and 2, *la*,) we see the hard, dark brown mandibles (Fig. 1, *md*,) having strong, toothed edges with which the grasshopper cuts off leaves of plants. Between these is the mouth. Below the mouth lies what is often called the under lip, but is really a pair of united appendages, the second pair of maxillæ (Fig. 1, mx^2). Above these and nearly hidden between them and the mandibles are the first pair of maxillæ (Fig. 1, mx^1). In Fig. 2 only the palpi or jointed feelers of the maxillæ are seen.

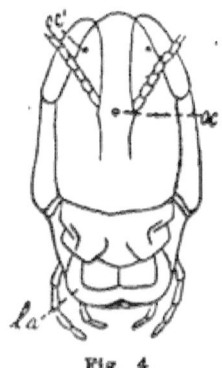

Fig. 4.

Fig. 3, *tn*, is the tongue, which is between the first pair of maxillæ.

Some of us have watched the grasshopper breathe, and know how he seems to pant as his body contracts and expands. We look on the sides of the abdomen for the breathing holes. Here they are, seven of them in plain sight, and another high up on the first ring in front of the ear (Fig. 1, C, s^3–s^{10}). They look like tiny pinholes, just above the fold on each side of the abdomen and close to the forward margin of each ring. If these do not show plainly, a larger pair on the thorax (Fig. 1,

B, s^2,) can often be seen a little above the second pair of legs. There is also one more pair on the first ring of the thorax.

Through the breathing holes air passes to the wonderful sets of air tubes and air sacs found in every part of the grasshopper's body. Fifty-three of these tiny balloons have been counted in the head alone. A body made so light and buoyant is easily carried through the air by the strong wings.

The baby grasshopper, called the larva, is like its mother, but has no wings. In order to grow it must throw off its outer coat, just as the young lobster does. After it has done this three times little wing-pads appear on its back, and it is now called a pupa. Twice more it casts off its coat, and now it is the full grown imago, ready for flying or jumping.

The note of the grasshopper is produced by rubbing the small teeth on the inner side of the thigh of the hind leg against the veins on the outer side of the fore wing, or wing-cover as it is often called. It is easy to show how this is done by drawing a comb over the edge of a piece of stiff paper.

THE CRICKET.

Most of our crickets are females (Fig. 1) collected by the children in October as they were laying their eggs by the roadsides or in gravelly walks. After studying the grasshopper, it is easy to begin work upon this insect, and natural to compare the two at every step.

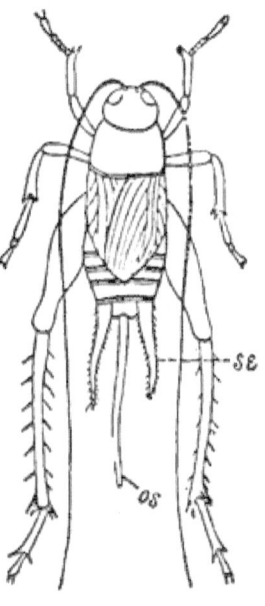

FIG. 1.

The cricket has a shorter and broader body than the grasshopper. The two sides of its body are alike. The body is in three parts,—head, thorax, and abdomen. It has one pair of very long antennæ. It has a pair of compound eyes, not so large as the grasshopper's. It has three pairs of legs on the thorax, the hind pair very long because it is a jumper. The forward wings, or wing covers, are short, covering only part of the abdomen. The hind wings are very small and of no use in flying. At the end of the abdomen there is a long bristle (Fig. 1. *se*) on each side. The egg-layer looks like a long sting (*os*).

Nine rings can be plainly counted on the back of the abdomen of the female. On each side of the abdomen is a soft space where the breathing-holes are seen. The ovipositor separates into two parts and each of these into two more, making four long, sharp piercers, which bore a hole in the ground and then unite to form a canal through which the eggs pass down into the earth.

The first ring of the thorax has a cape like the grasshopper's, but a straight, broad one, much more like a wide collar. The three rings are plainly seen on the under side. The hind legs are long but not so strong as the grasshopper's. The wing-covers are bent to fit the sides of the body. The wing-covers of the male are larger than those of the female, and have a different arrangement of the veins.

It is by rubbing these strong veins of the wing-covers together that the male cricket makes the lively chirp we know so well.

The head of the cricket is shorter and broader than that of the grasshopper, but like that is placed at right angles to the body. The eyes are not so large as the grasshopper's. The palpi are longer than the grasshopper's. Just below the upper lip the hard mandibles can be felt. With a pin the first pair of maxillæ with the long palpi can be pushed outward, and the united second pair of maxillæ with a shorter pair of palpi can be bent downward.

The strong mandibles are evidently fitted for biting, and we know that crickets do eat the tender parts of plants, even attacking roots and fruits. When abundant, they do great damage.

The eggs laid in the autumn are hatched the next summer. Most of the old insects die before cold weather comes, but a few live through the winter under stones or in dry holes.

The children must not fail to see and admire the white climbing cricket, daintiest and most delicately fashioned of all our New

The Cricket.

England crickets, which lives on trees and shrubs, but often seeks the shelter of our houses when the cooler nights come in early autumn. The male (Fig. 2) is white with a few yellow markings, with antennæ and legs so fragile and slender that a single careless touch will ruin a specimen. But let one perch on the window-sill for one night, and we see how effectually this delicate little creature can banish sleep. By rubbing together the three large oblique veins on the flat surface of the wing-covers he produces the loudest and shrillest of all cricket notes, and keeps up his serenade with unfailing energy and patience till daylight. The female, which has no note, is somewhat larger than the male, with narrower wing-covers and of a pale greenish or yellowish color.

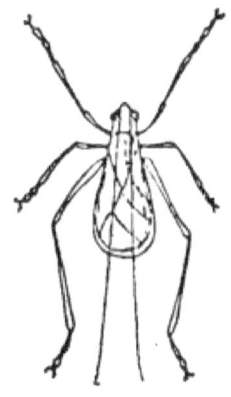

FIG 2.

THE BEETLE.

Specimens for this lesson are obtained on warm May or June evenings, when the blundering brown May-bug, June-bug, or dor-bug (Figs. 1 and 2), as it is variously called, enters at every window left open after the evening lamps are lighted, and in its headlong fashion goes bumping into anything and everything that stands in its way. It is said that they may also be collected by shaking the fruit trees, where they hide, at an early hour in the morning, when they do not attempt to fly, but fall to the ground. These beetles do so much harm in our gardens that we do not hesitate to put as many as we need into our bottle of alcohol. For our insect-boxes we pin beetles through the right wing-cover.

FIG. 1.

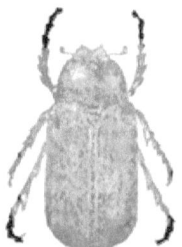
FIG. 2.

The children are all sure this is a bug; they have always heard it called so, and the word beetle is not found in their vocabulary. But Figs. 3 and 4 show that the bug has a long sucking-tube, which the beetle has not. This will be a sufficient distinction for the present, until true bugs are studied. Now we call our specimen the June-*beetle*.

What has the beetle that the grasshopper and the cricket have also?

It has the three parts of the body,—head, thorax, and abdomen. It has three pairs of legs. It has two pairs of wings. It has one pair of compound eyes.

Later we shall add that it has one pair of antennæ, but the children may not see them on this beetle till the head is carefully examined by itself.

The grasshopper, the cricket, and the beetle, are all called *insects*. In studying our new insect we will compare it with the grasshopper.

The body is shorter and broader than the grasshopper's. It is covered with a horny crust, much harder than the grasshopper's. This crust is its skeleton.

The fore wings are like two hard shells covering the back. They protect the other wings, and are called wing-covers. They meet in a straight line down the back, and cover the hind wings completely. If we should cut the beetle in two between the wing-covers, the two halves would be just the same size. There is a little shield between the wing-covers.

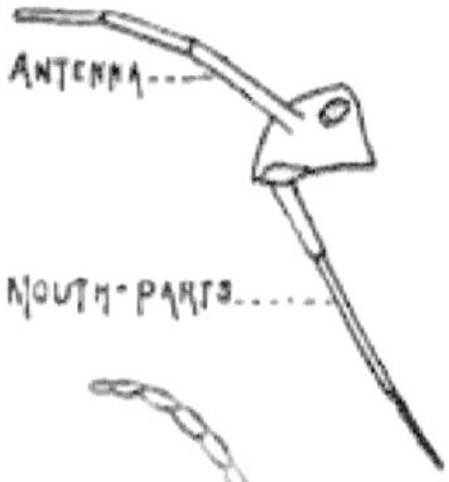

FIG. 3.
Mouth-Parts of Bug.

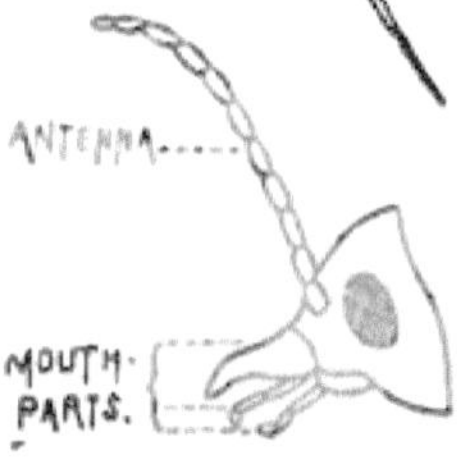

The first ring of the thorax is very large, the second and third, though large, are not seen on the back, with the exception of the little shield that belongs to the second ring.

FIG. 4.
Mouth-Parts of Beetle

The hind wings are thin with very strong veins and a joint near the middle so that they can be doubled up under the wing-covers.

The female uses the strong spines on the legs in digging her way into the earth, where she lays her eggs.

The broad and short abdomen is soft on the back because protected by the wing-covers. There is no egg-layer. The breathing-holes are plainly seen on the sides of the rings.

94 *Lessons in Zoölogy.*

The beetle has a pair of compound eyes. It has a little flat piece that comes out over the mouth. There is

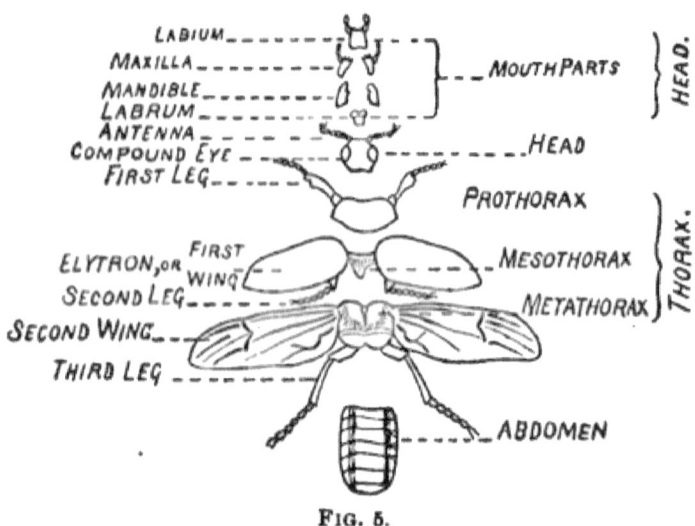

FIG. 5.

a pair of queer little things that have an elbow near the middle and a little club at the end. These must be the antennæ, and the little club is made of three leaves.

With magnifying glasses we also see at least one pair of palpi, and with a pin we assure ourselves that the beetle has hard mandibles.

From the card on which the mouth-parts are glued, and from the blackboard drawing of Fig. 4, the class observe that the beetle has the same mouth-parts as the grasshopper. In Fig. 5 the united second maxillæ are indicated by their other name of labium or lower lip.

On a male stag beetle (Fig. 6) we now find the enormous hooked mandibles, the long first pair and short second pair of palpi, the sort of tongue formed of brushes of hairs attached to

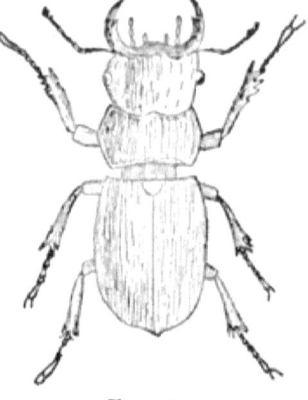

FIG. 6.

both pairs of maxillæ and the antennæ ending in four separated leaves.

The grub of the June beetle (Fig. 7) is a large white worm with a brownish head and strong jaws, which

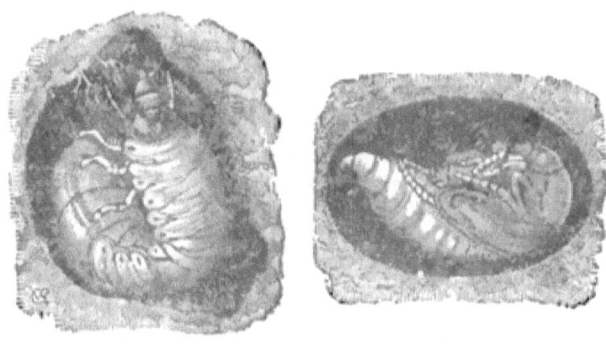

FIG. 7.   FIG. 8.

lives in the earth and devours the roots of grass and other plants. The pupa (Fig. 8) lies quietly in its cocoon in the earth until it comes out as the perfect insect.

THE DRAGON-FLY.

Dragon-flies may be caught with a net near ponds and streams, where the larvæ are found in the water during July and August, and the pupæ in spring and autumn. The larvæ and pupæ can be kept in the schoolroom in a jar of water, with sand at the bottom. The latter will need a water-plant, or some leaves and twigs, to which they can cling. The contrast between the sluggish young and the swift-flying adult insect will seem to children so marvelous that it will awaken a new interest in insect study. It gives them a glimpse of the possibilities the insect world offers of new discoveries, and makes them eager to observe living forms, the very thing we most want them to do.

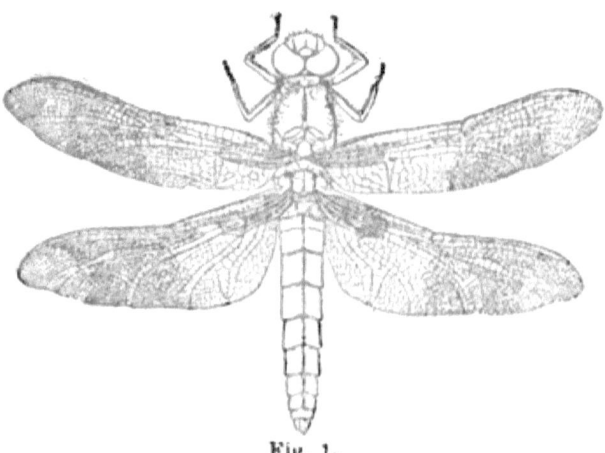

Fig. 1.

The dragon-fly has the three parts of the body, but the abdomen is very long. His head is so loose it looks as if it might drop off. He has a hump on his back because his thorax is so high. He has three pairs of legs and two pairs of wings. The legs are all crowded together. The wings are very large, and the two pairs are nearly the same size. He uses both pairs in flying. The dragon-fly has enormous compound eyes, that meet on the top of his head. His antennæ are only two little bristles.

The abdomen has ten rings. It has a very small egg-layer.

The horny ridges on the second and third rings, one of which is seen in Fig. 1, may lead pupils to count twelve rings instead of ten, but by comparison with the sutures between the rings, the ridges are seen not to be true sutures.

All the rings of the thorax can be seen on the back of the dragon-fly. The first ring is very small, but the second and third rings are large because they carry those great wings. The wings are beautifully veined. There are long, straight veins on the front margin, and the rest of the wing is net-veined.

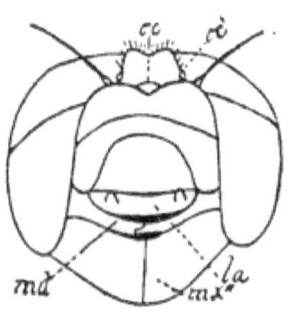
Fig. 2.

Fig. 3.

As the dragon-fly never alights, but always hangs by the second and third pairs of feet, the legs are drawn forward and the rings of the thorax inclined in the same direction. A voracious eater, the deadly enemy of gnats and mosquitoes, the dragon-fly must catch its food "on the fly." For this purpose, see its immense compound eyes, literally "on all sides of its head"; the head itself so loosely hung that it can be turned in any direction, or thrown backward till it touches the second ring of the thorax, while the first ring of the thorax moves so freely that the first pair of legs, used only for seizing prey, can readily follow the rapid motions of the head.

Grasping the head and thorax of the dragon-fly firmly, and looking at the head from in front, as shown in Fig. 2, we see a little horny projection in front of the compound eyes and between the antennæ. The largest of the simple eyes (Fig. 2, *oc*) is in front of this projection, the two smaller (*oc'*) at either side of it. The upper lip (*la*) is easily lifted with a pin, and just below it are the dark brown, horny mandibles (*md*) and the second maxillæ (*mx''*), which hide away in their broad cavity the small first maxillæ and the thick tongue.

Fig. 3 represents a larva collected in August, three times life size, *mx''* being the mask; *w'* and *w''*, wing-pads just appearing.

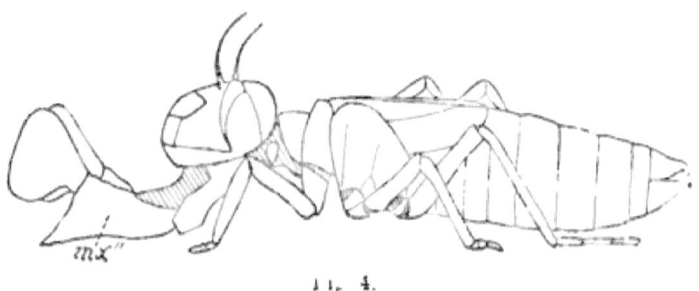

Fig. 4.

In Fig. 4, the pupa is shown with the spoon-shaped mask (*mx''*) extended to seize food. This mask is the greatly enlarged second pair of maxillæ, and when not in use, is folded back over the mouth so as to hide the strong mandibles.

The dragon-fly lays its eggs on water-plants. When the pupa is ready for the last change, it climbs up on some plant, the skin of its back splits open, and the dragon-fly pulls itself out. This insect does not deserve the name "darning-needle," since it has neither sting nor powerful jaws, but its threatening appearance may well gain for it the name dragon-fly.

THE BUG.

Most of us have a prejudice against *bugs*, which has some reason for its existence in the disagreeable character of many that bear this name, but the remarkable adaptations to their mode of life shown by some of these insects will awaken our interest in spite of ourselves.

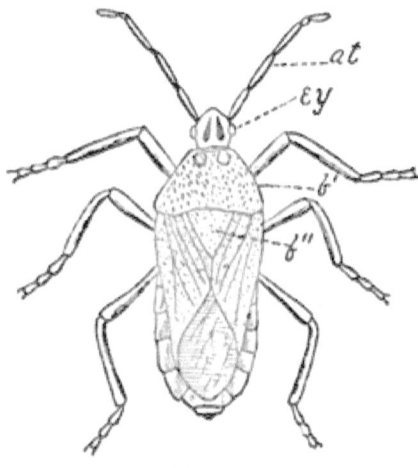

FIG. 1.

The squash-bug, shown enlarged in Fig. 1, gives a good idea of the characteristic form of the body, and also of the peculiar wings, and is easily found on squash-vines. It is too small for the mouth-parts, but these are well shown by the common Cicada, or harvest fly (Fig. 2), while if the teacher can have one of the "giant water-bugs" (Fig. 3), she will find it invaluable. One of the most interesting species, and therefore one that we cannot afford to do without, is the lively water-boatman (Fig. 4), which children can collect from the ponds.

The squash-bug has a small, pointed head. It has head, thorax, and abdomen. The abdomen is flat on the back and rounded below. The head is much lower than the back of the thorax and the abdomen. There are three pairs of legs, used for walking, not for jumping. There are two pairs of wings. The forward pair are thickened in front and thin behind, and they overlap. They are called wing-covers. The hind wings are thin.

Some will undoubtedly fail to see from their specimens that the wing-covers are half membranous, but the "giant" will make this very clear.

The abdomen has no appendages. It has a little rim spotted with yellow, that comes out beyond the wing-covers. The breathing-holes are on the sides.

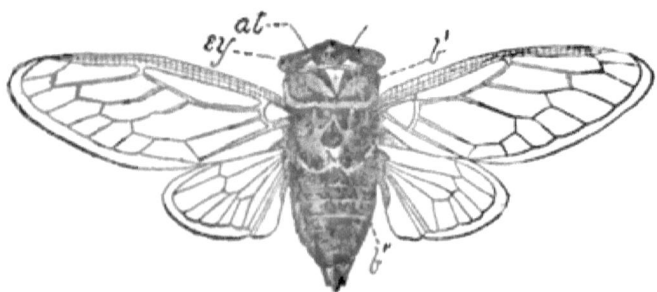

FIG. 2.

On the under side of the body the three rings of the thorax are seen, but only the first ring (Fig. 1, b') and the shield (b'') of the second on the upper side.

Besides the two rather small compound eyes, there are two simple eyes. The antennæ are long and rather large. The mouth parts look like a long sting.

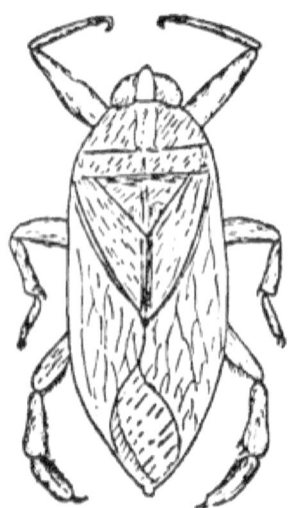

FIG. 3.

The mouth-parts of the Cicada are like those of the squash-bug magnified, and are large enough for children to separate and describe, so we turn to the Cicada now.

Fig 5 is a magnified view of the head of the squash-bug with the mouth-parts separated.

There are two pairs of long, sharp needles for piercing (Fig. 5, mx', md). There is a short, horny upper lip (lu). There is a long, jointed tube (mx''), in which the two pairs of needles lie.

FIG 4.

Since we have always found three pairs of mouth-parts on insects, and have already seen that the second maxillæ may be so changed as to form the mask of the young dragon-fly, children will see after a little questioning, if not before, that the long tube is the second pair of maxillæ, and the needles are modified mandibles and first maxillæ. The tissues of plants are pierced with the sharp needles,

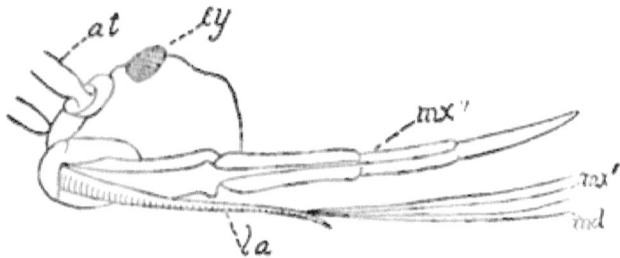

FIG. 5.

and their juices are drawn in through the sucking-tube. The triangular head, with its broad base, is firmly braced against the broad thorax to furnish a strong support for the thrust with the needles.

In August the light brown larva (Fig. 6, three times life-size) and the pupa of the squash-bug can be collected on the leaves of squash-vines. The pupa resembles the full-grown insect, but is lighter in color and has only wing-pads in place of wings.

Comparison of the beetle and the bug. — The beetle has a hard crust. Its wing-covers are hard and horny, and meet in a straight line down the back. The hind wings are doubled up under the wing-covers. The beetle has hard mandibles for biting, and two pairs of maxillæ with palpi.

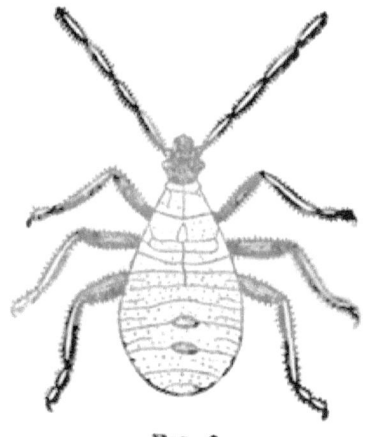

FIG. 6.

The bug has wing-covers half horny and crossing on the back. It has a small, pointed head. Its mouth-parts are a sucking-tube and two pairs of needles.

This comparison is sufficient to enable one always to distinguish these two orders of insects. If, however, we have observed them in all stages of growth, we can add the very important difference that while the young bug resembles its parent in a general way and the active pupa is still more like it, both living on squash-vines, the larva of the June beetle is worm-like and lives in the ground, and the pupa is quiet in a cocoon.

THE CICADA.

The Cicada, or harvest-fly, will furnish material for an interesting lesson. An abundance of the cast-off pupa skins can be collected by the children, and will not only show perfectly the appearance of the pupa, but will give an excellent idea of the horny external skeleton of an insect and the complete manner in which it is stripped off to allow for growth.

The body is broad and short. The eyes (Fig. 1, *ey*) stand out on the sides of the head, and there are three simple eyes between them. The antennæ (*at*) are like bristles. The first and second rings of the thorax (*b'* and *b''*) are very broad; the third is very narrow, because it carries only the small hind wings.

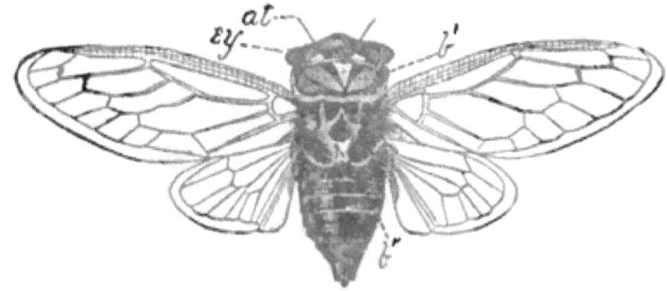

Fig. 1.

There are light spots on the head and thorax, and on the second ring of the thorax a marking that looks like the letter *W*. This mark was long supposed to stand for the word *War*, and made the superstitious believe the harvest-fly an insect of ill omen.

The wings slope like a roof over the sides of the body, and both pairs are thin. The veins of the fore wings are very large and strong. The abdomen ends in an egg-layer. Some of the harvest-flies have two broad plates on the under side of the abdomen.

These plates are found on the male, and cover the kettledrums, by which he makes the shrill sound we know so well in dog-days.

The mouth-parts having been described in the last lesson, are omitted here, but would of course be reviewed in the schoolroom.

The pupa has very large fore legs, like great claws. It has a sucking-tube like its parent. All the rings of the thorax show plainly. The first and second rings are very large; the third ring is small. The horny outer layer of the antennæ and the eyes is cast off with the rest of the skeleton.

Fig. 2.   Fig. 3.

With the strong, horny piercer at the end of the abdomen the female makes hollows in twigs in which to lay her eggs. The larva (Fig. 2) is hatched on the tree, but lets itself fall to the ground, being so light that it descends very slowly and is not injured by the fall. It then burrows in the earth, where it sucks the sap from roots. One species of harvest-fly passes seventeen years in the earth as larva and pupa; others common in New England require only one or two years for their transformations. The pupa at last digs its way out of the earth with its big fore claws, climbs some tree, its skin splits on the back, and the harvest-fly comes out.

Fig. 3 shows the insect making its way out of the pupa skin.

The water-boatman, mentioned in the last lesson, is a different type, and should be studied from living insects kept in jars in the schoolroom. The jars must be covered with netting to prevent these active little bugs from flying away.

We discover that the water-boatman is a true bug by feeding it with meat and watching the sucking-tube, or by handling it carelessly and letting it use its sharp needles on our fingers.

Its body is boat-shaped, its back being the keel, hence it swims back downward. The under side, which is flat, forms the deck, with an upper deck of hairs above it. Under the wings it keeps the supply of air for which it often comes to the surface. The fringed hind legs are the principal pair of oars, working smoothly in their rowlocks. The other two pairs are held out in front for seizing prey. It feathers its oars by pressing the hairs to the hind leg when drawing it forward, and spreading them in drawing it back.

THE FLY.

The big buzzing "blue-bottle," or any of the common flies found about houses, can be used for this lesson, though the green-headed horse-fly (Fig. 1) is better, if it can be obtained. Most of the descriptions here given will apply to either of these. The crane-flies with their delicate wings and long, slender bodies, are also invaluable as specimens.

The fly is an insect. It has a short, broad body. It has head, thorax, and abdomen. It has three pairs of legs. It has one pair of wings and one pair of balancers

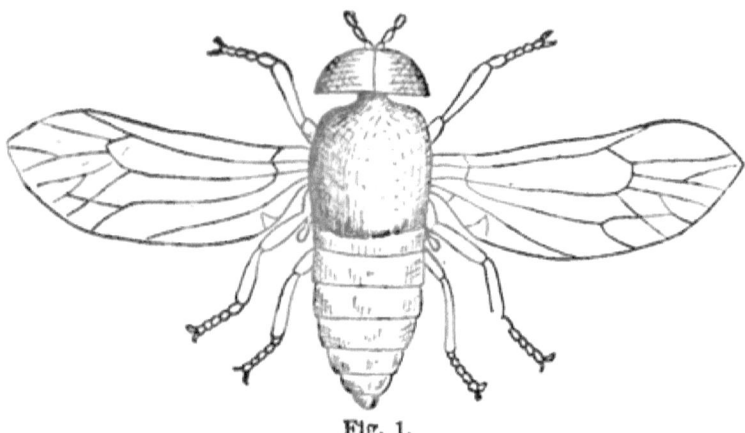

Fig. 1.

The last statement can be made only after careful observation. The little scale or winglet on each side (Fig. 2, *s c*) will at first be taken for another wing, but after drawing the wing forward several times we discover that the winglet moves with it and must be a part of it. Hence there can be but one pair of wings. Just under the winglets is a pair of tiny whitish knobs on slender stems (Fig. 3, *w″*) known as the balancers.

Fig. 2 represents the second ring of the thorax of the horse-fly, and Fig. 3 the third ring, with the appendages of one side. The balancers, being the greatly reduced second pair of wings, are borne on the third thoracic ring.

The Fly.

The fly has very large compound eyes that make its head look very broad. The house-fly has short, feathery antennæ. Its tongue is large and broad at the end.

The abdomen is very short and broad in front. It is covered with hairs. We can see only four or five rings.

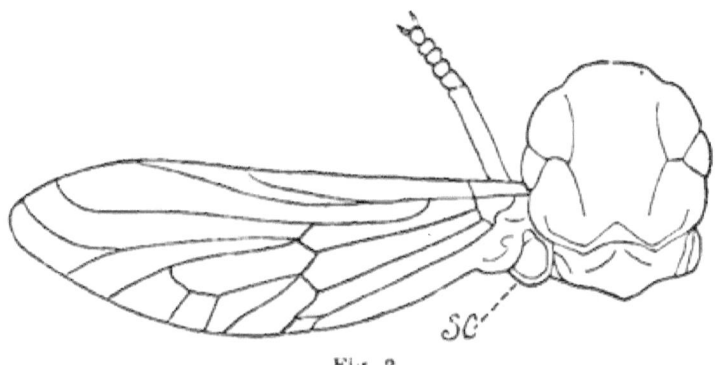

Fig. 2.

The thorax is nearly as large as the abdomen. It is covered with hair. It is joined to the head by a small neck. The legs are all nearly the same size. The feet have two claws and two little light-colored cushions.

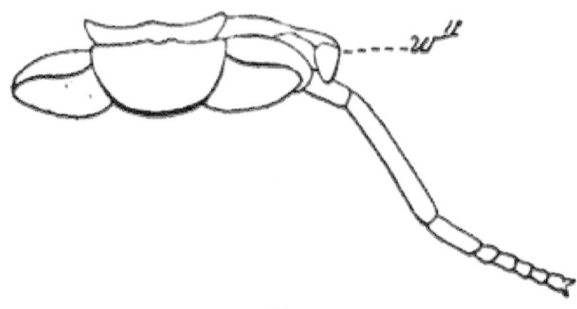

Fig. 3.

The little cushion, so deeply cleft that it looks like two, has on its surface hairs that pour out a sticky liquid by which the fly clings when walking upside down.

The antennæ of the house fly are usually in alcoholic specimens bent down closely against the face, but with the feathery bristles projecting more or less, so that one is apt to mistake them for the whole of the antennæ. They are really, however, attached to one side of the true feelers. The tongue, consisting of the modified second maxillæ, is the only conspicuous mouth-part. There are neither mandibles nor first maxillæ in this form, but with a good magnifying glass one can see the large palpi of the first maxillæ and the long, horny upper lip closely pressed over the upper side of the tongue. The two broad lobes that form the end of the tongue are roughened by cross-bars (Fig. 4).

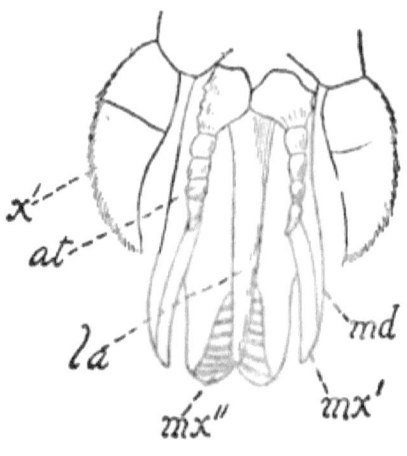

Fig. 4.

The house-fly simply laps his food, and has no need of pincers, but the horse-fly has sharp, lance-like mandibles (Fig 4, md) and first maxillæ (mx'), adapted for piercing the skin of horses and cattle, in addition to the long labrum, or upper lip (la), the broad tongue (mx''), and the palpi of the first maxillæ (x'). The antennæ (at) are also shown. In the curious large robber-flies, or insect-hawks (Fig. 5), which attack bees, beetles, and other insects, the mouth-parts form a powerful black sting.

The eggs of the house-fly are laid in stables, where the larva lives for several days as a white maggot without feet, then in a week of quiet the pupa changes to the perfect insect, which buzzes about our houses for a few weeks more.

In the crane-fly the balancers stand out well from the body, and one can see plainly that their slender stalks are borne on the last ring of the thorax. The large, high, second ring of the thorax, the long abdomen and long wings remind us of the dragon-fly, but the single pair of wings and the prominent balancers show us it can be nothing but a fly, even without an examination of the mouth-parts.

Fig. 5.

The large flies that mimic bees and are common around plants, are excellent subjects for study after the bee has been taken.

THE BUTTERFLY.

Whatever difficulty they may have in securing specimens for other lessons, children can always catch butterflies and caterpillars. Butterflies may be killed with chloroform or benzine, and the wings spread on a setting-board made by nailing two cleats lengthwise on a piece of wood with a space between them wide enough for the body of the insect. When dried, the butterflies may be pinned in a box with strips of cork glued inside the bottom. Caterpillars are put in boxes, fed with fresh leaves, and kept through their transformations, while chrysalids can be gathered in the autumn.

Fig. 1.

Cabbage butterflies (Fig. 1) are always plenty, and will do very well, if larger ones are not at hand. Specimens for study may be kept without pinning in boxes or envelopes. Two days before the lesson they should be placed upon thin paper over wet sand. This will soften them so that they can be handled without breaking easily. Butterflies kept in alcohol will be perfectly flexible, but will have no bright colors, and children will not feel that they are the real thing.

The cabbage butterfly has two pairs of very large wings. They are white with two black spots and a black patch on the forward ones, and one black spot on the hind ones.

The Butterfly.

The male has but one black spot on each fore wing.

If we rub the dust off the wings, they lose their color and are transparent. When the butterfly rests, it carries its wings erect over its back. Its body is very small for its wings. The rings show plainly on the abdomen. The head and thorax are hairy.

The hairs and scales must be rubbed off in order that the three regions of the body may be clearly seen.

The butterfly has a pair of antennæ shaped like clubs with long handles. It has two large compound eyes. It has a long tongue coiled up tightly under its face (Fig. 2, mx^1,). It has two little bunches of hairs standing up in front of its forehead (Figs. 1 and 2, x^2).

Fig. 2 represents the sucking tube and one of the palpi of the monarch or milk-weed butterfly.

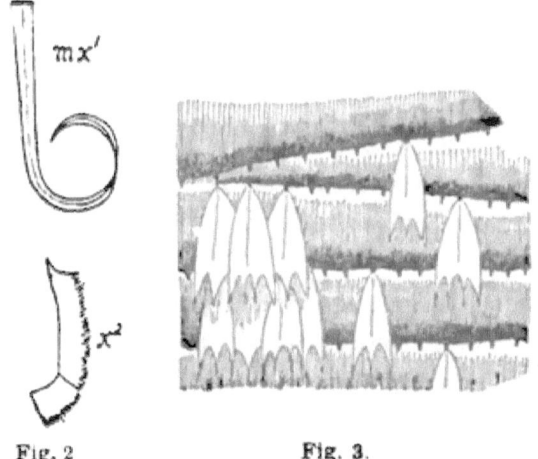

Fig. 2 Fig. 3.

The two "bunches of hairs" are the palpi belonging to the second pair of maxillæ. Though these maxillæ are either obsolete or very minute in butterflies, their palpi form two large hairy cushions, between which rests the coiled tongue or sucking tube, (Fig. 2, mx^1). The latter is the greatly lengthened first maxillæ, their edges united to form a closed tube, through which the honey of the flowers is carried to the mouth. Mandibles are of no use and extremely minute, if present at all.

After the scales are rubbed off, the children draw the wings, noticing the long veins evenly distributed over them. Such a wing cannot give a strong downward stroke, hence the fluttering of the butterfly. If a microscope can be procured, a bit of a wing with some of the scales still on it may now be shown (Fig. 3), each child looking at it in turn while the others draw.

The butterfly's legs are small and weak, because but little used.

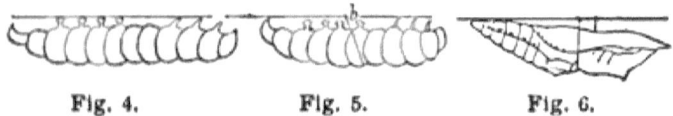

Fig. 4. Fig. 5. Fig. 6.

The caterpillar of the cabbage butterfly (Fig. 4) has a long, greenish body. The rings of the thorax are like those of the abdomen. The legs on the thorax are very small and end in little claws. There are five pairs of legs on the abdomen ending in broad cushions. The breathing holes show very plainly on the abdomen. Since the caterpillar feeds on leaves, it has strong mandibles for biting.

In its winter sleep in the chrysalis the caterpillar is transformed into a butterfly. It prepares for the change by seeking the under side of some fence rail (Fig. 4), fixing itself by its tail, and spinning a strong silken band around the middle of its body to hold it in place (Fig. 5). In its firm chrysalis skin (Fig. 6) it defies Jack Frost, and comes out in the spring a perfect butterfly.

THE MOTH.

This lesson may be made intensely interesting by using large moths and butterflies and comparing them at every step. It is not necessary that all the butterflies or all the moths should belong to the same species. The common milkweed butterfly, *Danais Archippus* (Fig. 1), or the large yellow and black swallow-tail, *Papilio Turnus*, will be excellent in comparison with the American silkworm moth, *Telea Polyphemus* (Fig. 2), or either of the large moths somewhat resembling it. If some of the class have hawk-moths (Fig. 6), it will be all the better, especially if they can tell from their own observation that these are more rapid fliers than the others, and thus see the use of the longer, more pointed, and more powerful fore wings and the small hind wings working with them. The Polyphemus is here taken first as the basis of the lesson.

Fig. 1.†

The body of the moth is broader and stronger than that of the butterfly. It has a thicker coating of hair. The wings are larger and not so brightly colored. The

---

† From Hyatt's *Insecta;* D. C. Heath & Co., publishers, Boston.

Fig. 2.†

† From Hyatt's *Insecta*; D. C. Heath & Co., publishers, Boston.

antennæ are not shaped like clubs, but are feathered. The mouth-parts are so small that nothing but thick cushions of soft hairs can be seen on the under side of the head.

This moth probably lives but a short time and eats little, since it can only lap up its food.

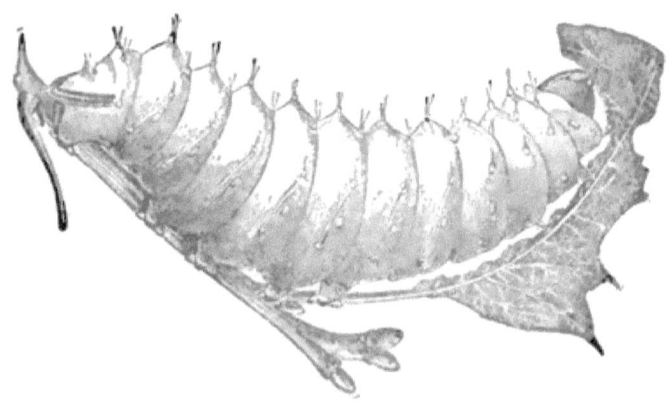

Fig. 3. †

Most moths fly at twilight or in the night, while butterflies fly by day. When moths are at rest, their wings form a sloping roof over the body.

The Polyphemus moth generally lays its eggs on oak leaves. The huge caterpillar (Fig. 3) is bright green *without yellow stripes or bands,* but with rows of hairy warts, and an oblique white line on the side of each ring. Its head and feet are brown, and the tail is bordered by a brown V-shaped line. It makes a beautiful cocoon of glossy silken threads covered on the outside with leaves (Fig. 4), in which it safely passes the winter. When the leaves fall in the autumn, the tough oval cocoon enclosed in them is borne to the ground. If one of these cocoons is opened, the pupa looks as in Fig. 5. Its body is much shorter than that of the caterpillar and covered with a

---

† From Hyatt's *Insecta;* D. C. Heath & Co., publishers, Boston.

hard brown skin. The wings and antennæ are glued to the under side of the body, and the breathing-holes show plainly on the sides of the rings. The moth comes out during May in Massachusetts. When the time comes for it to leave the cocoon, by means of an acid liquid it dissolves the gum that holds the silken threads together and then emerges without breaking the silk. Inside may be found the pupa-skin it has left behind.

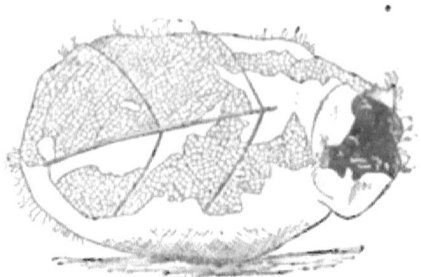

Fig. 4.

The hawk-moth (Fig. 6) has no bright-colored spots or markings. Its antennæ are not feathered and end in a small hook. Its sucking-tube is very long, so that it can reach the honey in long-tubed flowers. By scraping the hairs and scales from the under side of the wings, the little hook or bristle on the hind wing and the ring through which it passes on the fore wing can be seen.

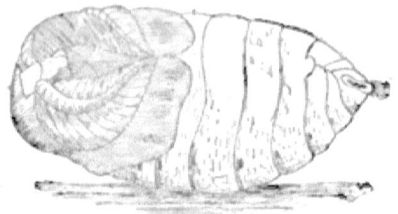

Fig. 5.

An interesting observation for pupils in the country is to watch this moth at twilight as it poises on its quivering wings over some large flower and thrusts its sucking-tube down to its base. From

The Moth.

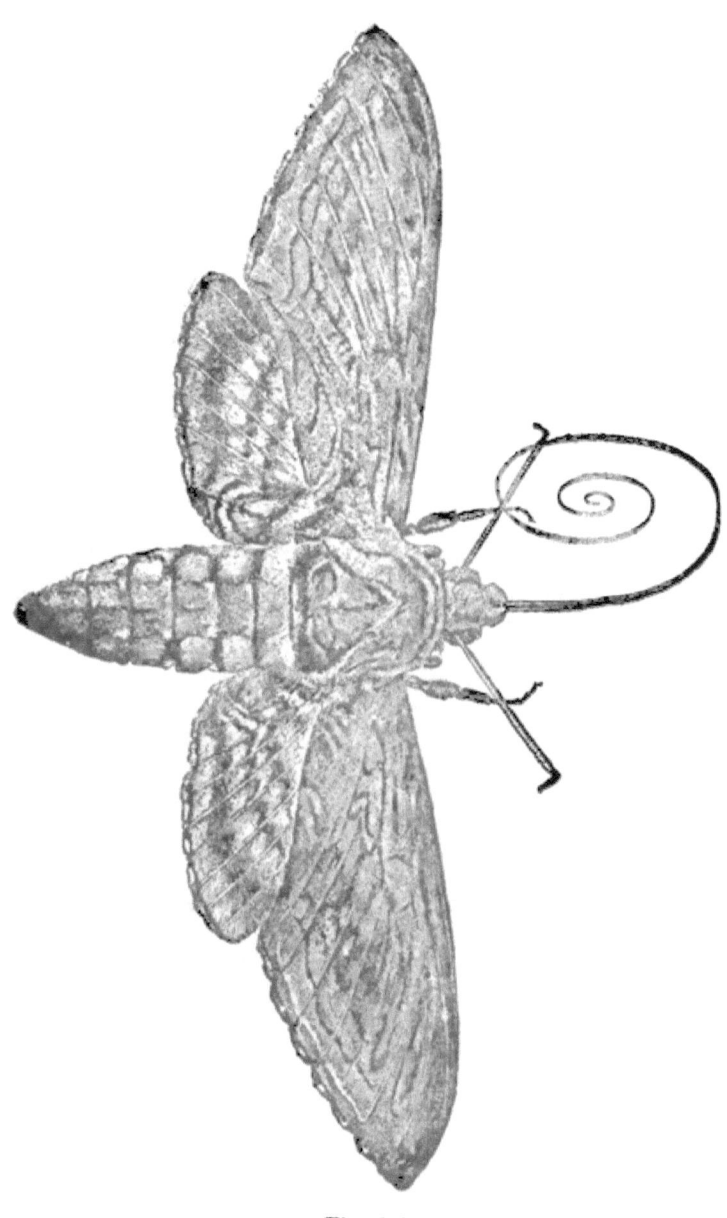

Fig 6 †

---

† From Hyatt's *Insecta*; D. U. Heath & Co., publishers, Boston.

its imitating the motions of the humming-bird in this way, it is often called the humming-bird moth.

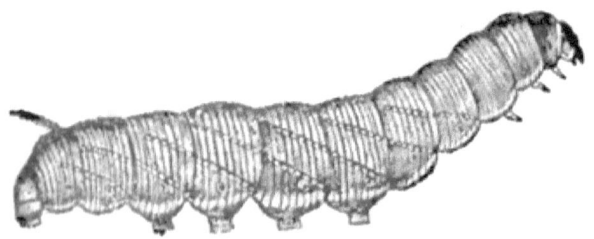

Fig. 7. †

The caterpillars of the large moths are all enormous eaters, and many of them are the great green "worms," so called. The "potato worm" (Fig. 7),—found also on tobacco and the tomato,—is the larva of the hawk-moth shown in Fig. 6. This larva will often remain for some time motionless on a stem with the head and front part of the body stretched upward, and from this habit the moths are named the Sphinxes.

---

† From Hyatt's *Insecta;* D. C. Heath & Co , publishers, Boston.

THE BEE.

Lesson I.

Of course we choose the honey-bee (Fig. 1) for this lesson. With a wide-mouthed bottle partly filled with dilute alcohol we take our stand beside a clump of tall plants in blossom, and in a single morning capture bees enough for a large class. When the busy little brown-coated worker has its head well buried in the flower, it is easy to place the bottle under the mouth of the flower and with the cork press the bee down into it. Only the knowledge that in no other way can we become acquainted with the structure of these little creatures reconciles us, however, to the murder of these industrious and useful insects.

Fig. 1.

The head and thorax are thickly covered with hairs. The body is short and strong. The three divisions of the body are plainly seen. The abdomen is connected with the thorax by a small joint. There are two pairs of wings, but the hind wings are so small and fit so closely to the edge of the fore wings that when they are spread they look like a single pair.

The hind legs (Fig. 2) are longer than the others, and the upper section of the foot (Fig. 2, f) is very broad. The lower section of the leg (Fig. 2, l) is concave on the inner side and surrounded by long hairs. The large section of the foot has several rows of stiff hairs across it. These broad joints are the "pollen baskets" in which pollen from the flowers is carried to the hive to be made into bee-bread for the babies. The foot ends in two claws.

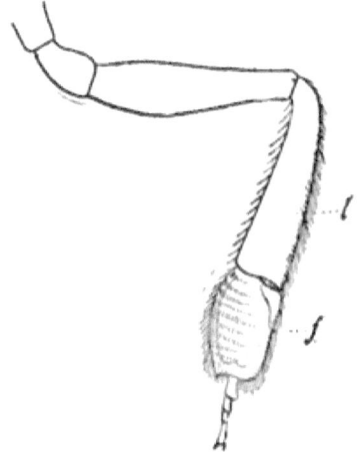

Fig. 2.

The few strong veins at the base and on the front margin of the wings enable them, though small, to strike the air with great force. The two wings on each side are so united by a series of hooks that they work together perfectly.

The antennæ (Fig. 3, *at*) are short. The compound eyes (Fig. 3, *ey*) are large. The bee has a pair of hard, brown mandibles (Fig. 3, *md*). Two pairs of light brown mouth-parts pointed at the end hang down below the mouth, besides a larger piece that looks like a tongue.

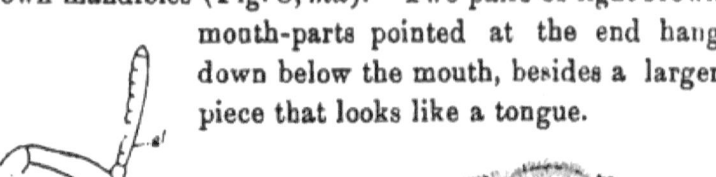

Fig. 3.

The shape of the head is one mark that distinguishes the bees from those flies that mimic them so well. In the fly the eyes always project further from the head. But if the difference in the shape of the head is not marked enough, the single pair of wings and the balancers of the fly will tell the story.

In the various insects we have studied we have seen mouth-parts for biting, for piercing, and for sucking, but here we find a combination of them all in one insect. The hard mandibles (Fig. 3. *md*) are perfect little organs for biting and cutting; the first maxillæ (Fig. 4, *mx.*) are long, sharp blades for piercing flowers; the middle portion of the second maxillæ (Fig. 4, *lg*) is the long sucking-

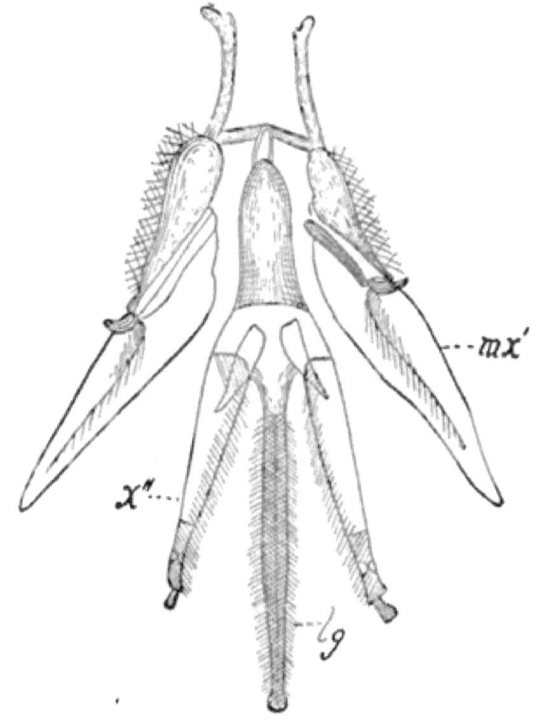

Fig. 4.

tube for gathering the nectar, the longest of all the mouth parts; and on either side of it is a long palp or feeler (x''), ending in tiny joints and reminding one of a brown needle. The best way to see the form and use of these parts is to capture a large bumble-bee and feed it with a syrup made of sugar and water.

The three simple eyes (Fig. 3, *oc*) do not show on alcoholic specimens unless they are allowed to dry, but are prominent on the living insect.

The sting is the egg-layer changed into a powerful weapon.

This lesson should be given in summer or early autumn, that the best part of it may be the observation of the living insect. We can follow single bees and see with what persistent industry they try all the flowers of one kind, paying not the slightest heed to any others, but so intent upon their work that it is easy to make them prisoners when once the sucking-tube is buried in a flower. Not the most brilliant buttercup can entice them if it is clover-honey that they are in search of. It has been proved beyond a doubt that bees can distinguish colors, and Sir John Lubbock decides, as the result of his experiments, that their favorite is *blue*.

Lesson II.

Some honey and a piece of the comb consisting of worker cells, drone cells, and a single royal cell; a queen bee and a drone, each in its vial of alcohol, and a few of the little worm-like larvae in another vial, — these constitute a perfect equipment for our second lesson on the bee. By making arrangements with a bee-keeper some months in advance, one may hope to secure all these, if she is fortunate.

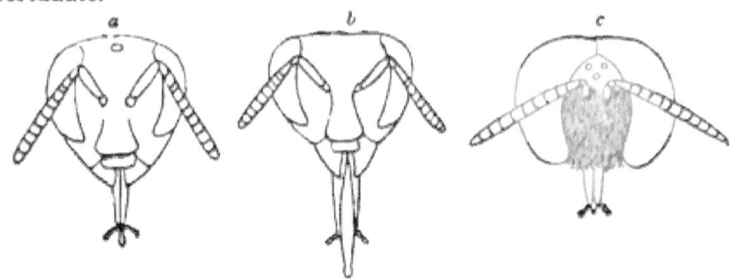

Fig. 1.

Many insects make homes for themselves or for their young. Caterpillars roll up leaves; the young clothes-moth hides itself in a case of fibers from our softest woolen dresses; the caddis-worm glues together a movable fortress of sticks and stones or a mossy covering of leaves, in which it conceals its greedy appetite and ugly jaws; and many kinds of bees and wasps tunnel stems or burrow in the earth, but one of the most remarkable insect homes is the honeycomb of the hive bee. Before examining the home, however, we must make the acquaintance of the other members of the family.

Fig. 2.

We have studied only the workers; but every colony must also have a queen, and during part of the summer, some drones. The queen, or mother bee, does nothing but lay eggs, sometimes as many as two or three thousand in a day. The drones, or males, do absolutely no work. The workers build the combs, gather the honey and pollen, act as nurses for the young, and attendants upon the queen.

We examine first the queen, then the drone, noting only distinguishing characteristics of each as compared with the workers, afterward observing the cells of the comb.

The queen is larger than the worker. Her head is

narrower than the thorax. The wings are shorter than the body of the queen. The lower joint of her hind legs is flat and has no fringe of hairs around it. It could not be used for a pollen basket.

This bee is a male, called a drone because he does no work. His head is narrower than the thorax, like the queen bee's. His eyes are not at all like the queen's or the worker's but meet on the top of his head, like a fly's. His wings are longer than his body. He has no pollen basket, but a very large upper joint on each hind foot.

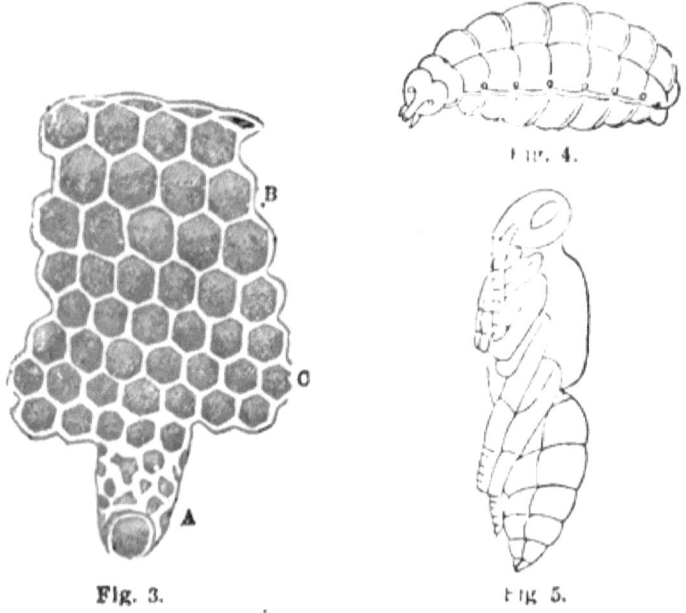

Fig. 3.   Fig. 4.   Fig. 5.

A colony may contain from twenty thousand to fifty thousand workers, and in summer from one thousand to two thousand drones, but never more than one queen at a time. When a new queen comes out of her cell, the bees "swarm," that is, the old queen leads a part of them away to form another colony. Or, if two queens do escape from their cells at the same time, a duel takes place which ends only with the death of one of the rivals.

In taking possession of a new hive they cling to one another in living festoons from the roof, while the wax forms in little plates under the edge of each ring of the abdomen. (Fig. 2.) These plates they take off with their feet, work them with their mandibles, and stick them to the roof of the hive till a shapeless mass of wax is formed. Then other bees hollow out the cells in the wax.

Fig. 1 represents the magnified head of each of the three different forms of the honey-bee; *a*, the queen, *b*, the worker, *c*, the drone, showing the great development of the mouth-parts in the worker and their very small size in the drone; also the differing shape of the face in queen and worker, as well as the enormous eyes of the male and the shortness of his antennæ above the bend.

Fig. 2 is a magnified view of a bee with the plates of wax showing between the rings of the abdomen.

In this piece of comb there are three kinds of cells: Small cells in rows, larger ones in rows, and a long one at the edge of the others. The cells in the rows are six-sided; the long one is shaped something like a peanut. The small cells (Fig. 3, *c*) are for the smallest bees, the workers; the larger ones (*b*) are for the drones; and the long cell (*a*) is for the queen. So few queens are produced every year that only a very few royal cells are needed. There are also storage cells for honey, which are sealed up with wax so as to be air-tight and keep the honey in its natural condition.

The larva (Fig. 4) is a tiny white worm, without feet, perfectly helpless, and entirely dependent upon its faithful nurses. In five or six days it reaches its full size, and covers itself with a thin cocoon. After ten days of pupa life (Fig. 5) it emerges from its cell, a perfect insect.

THE ANT.

The large black ant, that makes its nests in trees and may be seen around the outside of our houses, can often be easily collected in large numbers. The winged females (Fig. 1), appearing with the winged males (Fig. 2) in summer, are especially large, and the workers and soldiers make very good specimens.

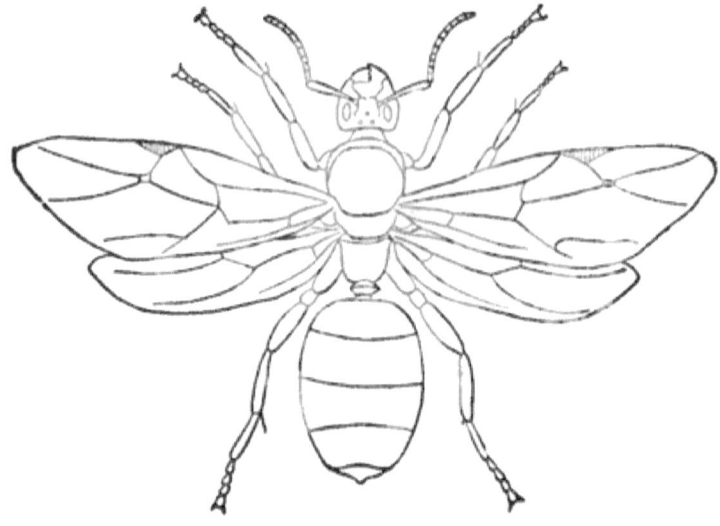

Fig. 1.

No better introduction to this lesson can be desired than to find an ant's nest in some rotten stump where the decaying bark and wood on the outside may be easily broken off, revealing the chambers within. What hurrying and scurrying as the ants rush back and forth carrying cocoons and helpless little grubs to safe retreats in the center of the nest! Utterly regardless of self and apparently incapable of fatigue, they will work for hours, if necessary, to place every one of their helpless charges under shelter. How they will tug and pull to rescue a cocoon that has been pinioned by a falling timber in the shape of a chip! How carefully they pick up the grubs with their sharp jaws, handling them so skillfully that the soft little creatures are never hurt! Whether they feel responsible for their charges or not, nothing could exceed the faithfulness of their care.

Our specimens are chiefly the common wingless forms,—workers (Fig. 3) and soldiers (Fig. 4). Placing these two kinds side by side, these observations will be made:—

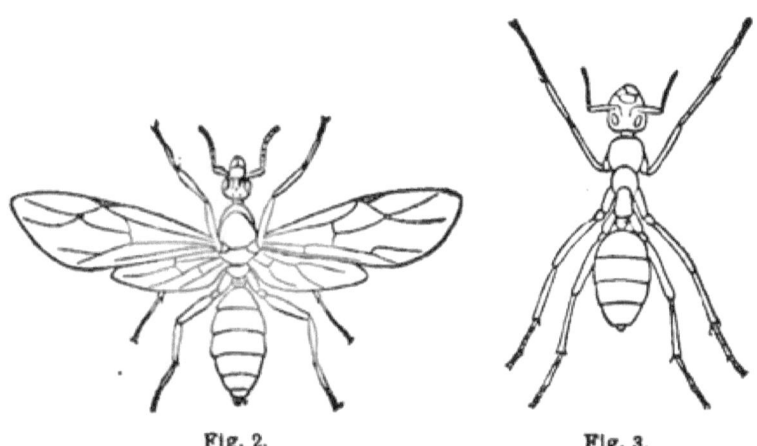

Fig. 2. Fig. 3.

The ant has no wings. Some of the ants have enormous heads. The head looks nearly as large as the abdomen. The thorax is very long and narrow. I can see the three rings of the thorax and a pair of legs on each ring. The abdomen can be curled up under the thorax. The jaws are very large. Two of the ants are all snarled up together; they have caught hold of each other's legs with their jaws, and it is almost impossible to pull them apart. Some of them are larger than the others, with longer bodies, larger heads, and stronger jaws.

As is shown by the figures, the soldiers are the larger forms. Their mandibles are enormously developed for use as weapons. Though the jaws of the workers are smaller, they are very strong, the ants carrying with them objects even larger than themselves for considerable distances.

Careful examination of the stem, or peduncle, joining the thorax to the abdomen, shows that in this ant it consists of two joints, making the abdomen more flexible and capable of being used with greater force in stinging.

The antennæ are bent near the middle. The head is longer and more pointed than the bee's, and the eyes are small. The head is triangular.

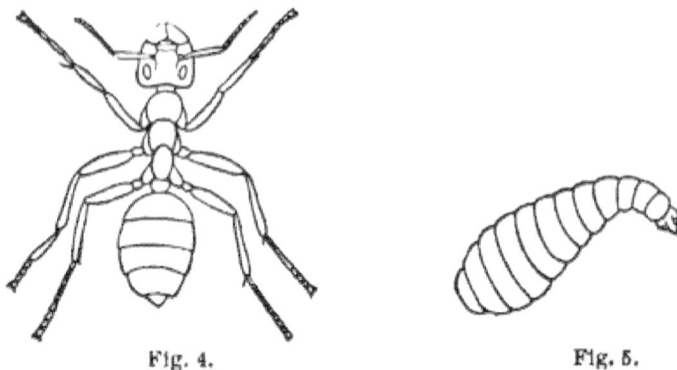

Fig. 4.   Fig. 5.

The males and females have three simple eyes (Figs. 1 and 2); the workers and soldiers have only the compound eyes. To see the mouth-parts remove the head from the body and hold it firmly, with sharp-pointed forceps, under a dissecting microscope, then with a needle pry apart the mandibles and carefully separate the parts below them from the head. The soft mass thus removed will be found to consist of a pair of broad, leaf-like first maxillæ, with long palpi, and the united second maxillæ, also with long palpi and with their inner side forming a thick, soft tongue. These parts can be glued to a card. This dissection is too difficult for children, however.

A few winged males and females should be shown, and the large size of the female noticed. The males die after the marriage flight, and the females pull off their own wings and settle down to a quiet life of several years, perhaps, in the nest. Sir John Lubbock had two queens that lived for seven years, and some of his workers lived six years.

The larvæ are small white grubs (Fig. 5), without legs, and so helpless that they cannot even feed themselves. They are dependent upon their nurses till they are full-grown, for many of them would die in the attempt to free themselves from their cocoons if not assisted by the older ants.

There is almost no limit to the number of interesting facts about ants that we can gather from such books as Sir John Lubbock's *Ants, Bees, and Wasps*, Mrs. Treat's *Chapters on Ants*, and McCook's *Agricultural Ant of Texas*. A few carefully chosen facts from such books, given to pupils after the observation lesson, will make them more eager to watch ants for discoveries of their own.

www.ingramcontent.com/pod-product-compliance
Lightning Source LLC
Chambersburg PA
CBHW020107170426
43199CB00009B/437